Electromigration Inside Logic Cells

Gracieli Posser • Sachin S. Sapatnekar
Ricardo Reis

Electromigration Inside Logic Cells

Modeling, Analyzing and Mitigating Signal Electromigration in NanoCMOS

 Springer

Gracieli Posser
Instituto de Informática - PPGC/PGMicro
Universidade Federal do Rio Grande do Sul
  (UFRGS)
Porto Alegre, Rio Grande do Sul, Brazil

Ricardo Reis
Instituto de Informática - PPGC/PGMicro
Universidade Federal do Rio Grande do Sul
  (UFRGS)
Porto Alegre, Rio Grande do Sul, Brazil

Sachin S. Sapatnekar
Department of Electrical and Computer
  Engineering
University of Minnesota
Minneapolis, MN, USA

ISBN 978-3-319-84041-3   ISBN 978-3-319-48899-8   (eBook)
DOI 10.1007/978-3-319-48899-8

Printed on acid-free paper

This Springer imprint is published by Springer Nature
The registered company is Springer International Publishing AG
The registered company address is: Gewerbestrasse 11, 6330 Cham, Switzerland

Preface

Electromigration (EM) in on-chip metal interconnects is a critical reliability-driven failure mechanism in nanometer-scale technologies. Usually, works in the literature that address EM are concerned with power network EM and cell to cell interconnection EM. This work deals with another aspect of the EM problem, the cell-internal EM. This work specifically addresses the problem of electromigration on signal interconnects and on Vdd and Vss rails within a standard cell. There are few studies in the literature addressing this problem, and the ones that can model EM are using a very simple model. To our best knowledge, this is the first work that analyzes, models, and reduces the EM effects on the signals inside cells by projecting the pin placement. In this work, cell-internal EM is modeled incorporating Joule heating effects and current divergence and is used to analyze the lifetime of large benchmark circuits. An efficient graph-based algorithm is developed to speed up the characterization of cell-internal EM. This algorithm estimates the currents when the pin position is moved avoiding a new characterization for each pin position, producing an average error of just 0.53 % compared to SPICE simulation. A method for optimizing the output, Vdd, and Vss pin placement of the cells and consequently to optimize the circuit lifetime using minor layout modifications is proposed. To optimize the TTF of the circuits, just the LEF file is changed avoiding the critical pin positions; the cell layout is not changed. The circuit lifetime could be improved up to 62.50 % at the same area, delay, and power because changing the pin positions affects very marginally the routing. This lifetime improvement is achieved just avoiding the critical output pin positions of the cells, 78.54 % avoiding the critical Vdd pin positions, 89.89 % avoiding the critical Vss pin positions, and up to 80.95 % (from 1 to 5.25 years) when output, Vdd, and Vss pin positions are all optimized simultaneously. We also show the largest and smallest lifetimes over all pin candidates for a set of cells, where the lifetime of a cell can be improved up to 76× by the output pin placement. Moreover, some examples are presented to explain why some cells have a larger TTF improvement when the output pin position is changed. Cell layout optimization changes are suggested to improve the lifetime of the cells that have a very small TTF improvement by pin placement. At circuit level, we present an analysis of the EM effects on different metal layers and different wire lengths for signal wires (nets) that connect cells.

Porto Alegre, Rio Grande do Sul, Brazil   Gracieli Posser
Minneapolis, MN, USA   Sachin S. Sapatnekar
Porto Alegre, Rio Grande do Sul, Brazil   Ricardo Reis

Acknowledgments

This work was partially supported by the Brazilian National Council for Scientific and Technological Development (CNPq, Brazil), Coordination for the Improvement of Higher Education Personnel (CAPES), and the Graduate Programs in Computer Science (PPGC) and on Microelectronics (PGMICRO), Institute of Informatics, Universidade Federal do Rio Grande do Sul (UFRGS) in Brazil.

We thank the University of Minnesota and two other researchers: Vivek Mishra (from the University of Minnesota) and Palkesh Jain (from Qualcomm India).

Contents

List of Figures

List of Tables

Abbreviations

A	Ampere
a	Atto
AC	Alternating current
AOI	Complex logic gate composed by the AND, OR and inverter gates
ASIC	Application-specific integrated circuit
CAD	Computer-aided design
CMOS	Complementary metal-oxide semiconductor
DC	Direct current
F	farad
f	femto
FF	Flip–flop
GDS	Graphic data system
GND	Ground/negative supply voltage
HDL	Hardware description language
IC	Integrated circuit
INV	Inverter gate
J	Current density
LEF	Library exchange format
μ	Micro
m	Milli
MTTF	Mean time to failure
n	Nano
NAND	Logic gate that represents the Boolean function $(A \cdot B)$
NDR	Non-default routing rules
NLDM	Nonlinear delay models
NMOS	N-type metal-oxide semiconductor
NOR	Logic gate that represents the Boolean function $(A + B)$
Ω	Ohms
p	Pico
RMS	Root mean square
RTL	Register-transfer level

SDC	Synopsys design constraints
SPICE	Simulation program with integrated circuit emphasis
TTF	Time to failure
UFRGS	Universidade Federal do Rio Grande do Sul
VDD	Positive supply voltage

Chapter 1
Introduction

Electromigration (EM) is a major source of failure in on-chip metal interconnects and vias and is becoming a progressively increasing concern as technology feature sizes shrink.

EM is initiated by current flow through metal wires and may cause open-circuit failures over time in copper interconnects. Traditionally, EM has been a significant concern in global power delivery networks, which largely experience unidirectional current flow. However, EM analysis can no longer be restricted just to global wires. Traditional EM analysis has focused on higher metal layers, but with shrinking wire dimensions and increasing currents, the current densities in lower metal layers are also now in the range where EM effects are manifested.

Usually works in the literature that address EM are concerned with power network EM and cell to cell interconnection EM. This work deals with another aspect of the EM problem, the cell-internal EM. This work specifically addresses the problem of electromigration on signal interconnects and on Vdd and Vss rails within a standard cell. Where there are few studies in the literature addressing it. Domae and Ueda (2001) found void formed due to electromigration in the interconnection portion in a CMOS inverter and then proposes some ideas to reduce the current through the wire segments where the voids were formed. Jain and Jain (2012) cite that the standard-cell-internal-EM should be checked and the safe frequency of the cells at different operating points must be modeled. Commercial EDA characterization tools like Cadence Virtuoso Liberate Characterization Solution (Cadence 2016) and Synopsys SiliconSmart (Synopsys 2016) both can model for cell-internal EM. Although, their model is very simple using traditional safe frequency versus operating load.

No previous work has approached cell-internal EM from pin placement perspective. In this way, our work is the first one to use the pin placement to reduce the EM effects inside of the cells and this is the main contribution of this manuscript - projecting pin placement as a knob for reliability tradeoffs.

© Springer International Publishing AG 2017
G. Posser et al., *Electromigration Inside Logic Cells*,
DOI 10.1007/978-3-319-48899-8_1

In this work, cell-internal EM is modeled incorporating Joule heating effects and current divergence and is used to analyze the lifetime of large benchmark circuits. An efficient graph-based algorithm is developed to speed up the characterization of cell-internal EM. This algorithm estimates the currents when the pin position is moved avoiding a new characterization for each pin position, producing an average error of just 0.53 % compared to SPICE simulation. While the runtime is reduced by about 12 times, on average, as runtime improvement is dependent on the number of different pin candidates available in each cell. A method for optimizing the output, Vdd, and Vss pin placement of the cells and consequently to optimize the circuit lifetime using minor layout modifications is proposed.

To optimize the TTF of the circuits just the LEF file is changed avoiding the critical pin positions, the cell layout is not changed. The circuit lifetime could be improved up to 15.77× at the same area, delay, and power because changing the pin positions affects very marginally the routing. This lifetime improvement is achieved just avoiding the critical output pin positions of the cells, 81.73× avoiding the critical Vdd pin positions, 160.66× avoiding the critical Vss pin positions and up to 160.66× (from 0.02 to 2.51 years) when output, Vdd, and Vss pin positions are all optimized simultaneously.

We also show the largest and smallest lifetimes over all pin candidates for a set of cells, where the lifetime of a cell can be improved up to 76× by the output pin placement. Moreover, some examples are presented to explain why some cells have a larger TTF improvement when the output pin position is changed. Cell layout optimization changes are suggested to improve the lifetime of the cells that have a very small TTF improvement by pin placement. At circuit level, we present an analysis of the EM effects on different metal layers and different wire lengths for signal wires (nets) that connect cells.

1.1   Reliability and Electromigration

Reliability is a key issue in integrated circuits (IC) designs because users expect failure-free operation throughout the product's lifetime (Kludt et al. 2014; Li et al. 2015a; Reis et al. 2015). However, failure rates will likely grow as transistors and wires shrink and the supply voltage scales slowly, leading to higher current densities and temperatures. As a result, transistor and wire degrade faster shortening the product's lifetime (Abella et al. 2008; Patel 2014).

Nanometer CMOS reliability issues can be categorized into spatial and temporal effects as Fig. 1.1 illustrates.

Spatial effects are immediately visible right after production and are fixed in time and can be random (e.g., random dopant fluctuations (RDF), line edge roughness (LER), etc.) or systematic (e.g., gradient effects, etc.). Temporal effects, on the other hand, are time-varying and change depending on operating conditions such as the operating voltage, temperature, switching activity, presence, and activity of neighboring circuits. Temporal effects can be aging effects (e.g., electromigration

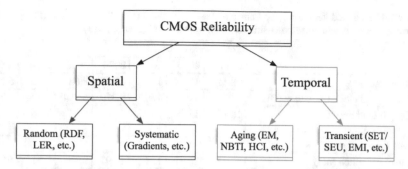

Fig. 1.1 A CMOS circuit can fail from spatial or temporal unreliability effects (Maricau and Gielen 2013; Wirth and da Silva 2010)

(EM), hot carrier injection (HCI), negative bias temperature instability (NBTI), etc.) (Sengupta and Sapatnekar 2014) and transient effects (e.g., single event transients (SETs), single event upsets (SEUs), electromagnetic interference (EMI), etc.) (Maricau and Gielen 2013). In this work, we investigate the electromigration effects which are one of the most important and challenging aging effects on integrated circuits, specially in nanometer technologies due to the increasing interconnect density, number of layers, and power consumption.

Electromigration (EM) is one of the critical reliability concerns, Wu et al. (2012) causing shorts and opens in metal interconnects, leading to interconnection failures and decreasing the mean time to failure (MTTF) of the chip. Particularly for copper metallization at 45 nm and below, EM affects global and local interconnects and is a major source of wire and via failure (Srinivasan et al. 2004) in a chip, limiting the performance scaling (Geden 2011; Kahng et al. 2013a; Vaidyanathan et al. 2014b; Xie et al. 2012). The gap between what circuit design needs and what technology allows is rapidly widening for maximum allowed current density in interconnects. This is the so-called EM crisis (Li et al. 2014). In this way, the concern of EM reliability has attracted more attention from circuit and chip designers, integrators, and reliability engineers (Li et al. 2014). Thereby, it challenges the state of the art in design, physics, process, and CAD which are dealing with EM to make new technology nodes feasible and reliable (Park et al. 2010).

1.2   Electromigration in Future Technologies (Lienig 2013)

With the technology miniaturization, line widths will continue to decrease over time, as well the wire cross-sectional areas. Table 1.1 (Lienig 2013) shows that the cross-sectional area shrinks from about $1000\,nm^2$ in 2014 to less than $500\,nm^2$ in 2018 (Lienig 2013). The currents are also reduced due to lower supply voltages and shrinking gate capacitances.

Table 1.1 Predicted technology parameters based on the ITRS, 2011 edition (ITRS 2011); maximum currents and current densities for copper at 105 °C (Lienig 2013)

	Year						
	2014	2016	2018	2020	2022	2024	2026
Gate length (nm)	18.41	15.34	12.78	10.65	8.88	7.40	6.16
On-chip local clock frequency (GHz)	4.211	4.555	4.927	5.329	5.764	6.234	6.743
DC equivalent maximum current (μA)[a]	18.14	12.96	10.33	7.36	5.53	4.45	3.52
Metal 1 properties							
Width—half-pitch (nm)	23.84	18.92	15.02	11.92	9.46	7.51	5.96
Aspect ratio	1.9	2.0	2.0	2.0	2.1	2.1	2.2
Layer thickness (nm)[a]	45.49	37.84	30.03	23.84	19.87	15.77	13.11
Cross-sectional area (nm^2)[b]	1079.7	716.0	451.0	284.1	187.9	118.4	78.13
DC equivalent current densities (A/cm^2)							
Maximum tolerable current density (w/o EM degradation)[b]	4.8	3.0	1.8	1.1	0.7	0.4	0.3
Maximum current density (solutions unknown)[b]	25.4	15.4	9.3	5.6	3.4	2.1	1.2
Required current density for driving four inverter gates	1.68	1.81	2.29	2.59	2.94	3.76	4.50

All remaining values are from the ITRS 2011 edition (ITRS 2011)
[a]Calculated values, based on given width W, aspect ratio A/R, and current density J in (ITRS 2011), calculated as follows: layer thickness $T = A/R \cdot W$, cross-sectional area $A = W \cdot T$ and current $I = J \cdot A$
[b]Approximated values from the ITRS figure INTC9 (ITRS 2011)

However, as current reduction is constrained by increasing frequencies, the decrease in cross-sectional areas increases the current densities J (as Fig. 1.2 shows). And, consequently decreases the lifetime by half each new generation, as presented in Fig. 1.3, where a critical area to EM is visible below about 28 nm. The green line in Fig. 1.3 shows the EM enhancement urgently needed what is aimed for advanced technology nodes.

According to Fig. 1.2 based on the Interconnect Chapter of the 2011 International Technology Roadmap for Semiconductors (ITRS) (ITRS 2011), the maximum current density limits (from EM reliability considerations) become the barrier to further frequency scaling from 2018 onwards. Moreover, the ITRS indicates that all minimum-sized interconnects will be EM-affected by 2018, potentially limiting any further downscaling of wire sizes (Fig. 1.2, yellow barrier) (Lienig 2013). As the total length of interconnect per IC will continue to increase, reliability requirements per length unit of the wires need to increase in order to maintain overall IC reliability. The ITRS thus states that no known solutions are available for the EM-related reliability requirements that we will face approximately in 2023 (Fig. 1.2, red barrier) (Lienig 2013). This shows the great importance of the research seeking for solutions that mitigate the EM effects mainly on the latest technologies.

Design tools can significantly improve the EM robustness of the generated layout by utilizing EM-optimized layout configurations as constraints during synthesis

Fig. 1.2 Expected development of current densities (J_{max}) needed for driving four inverter gates, according to ITRS 2011 (see also Table 1.1). EM degradation needs to be considered when crossing the *yellow barrier* of current densities (J_{EM}). As of now, manufacturable solutions are not known in the *red area* (Lienig 2013)

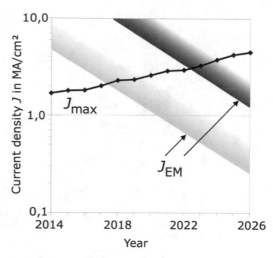

Fig. 1.3 Evolution of lifetime versus technology node. *Black line* shows the effect of reduced critical void volume: *Green line* shows the EM enhancement urgently needed (ITRS 2011) (*Courtesis of A. Aubel/ Globalfoundries*)

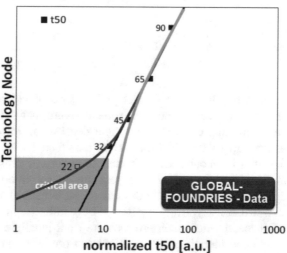

steps, such as routing. It is believed that this inclusion of EM-specific requirements in the physical design can provide relief from severe reliability constraints in future technologies (Lienig 2013).

1.3 Motivation and Contributions

Traditionally, EM has been a significant concern in global power delivery networks (Xie et al. 2012), where the direction of current flow is generally unidirectional, resulting in a steady migration pattern over time (Sapatnekar 2013). Recently, two new issues have emerged, as Fig. 1.4 shows.

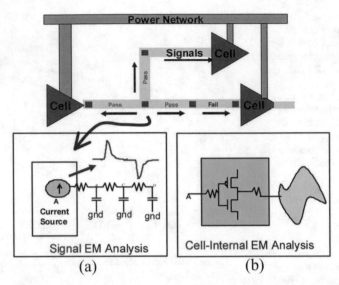

Fig. 1.4 Problem space: (**a**) current source modeling for signal-EM analysis and (**b**) load abstraction for cell-internal EM analysis (Jain and Jain 2012)

First, EM analysis can no longer be restricted just to global wires. Traditional EM analysis has focused on higher metal layers, but with shrinking wire dimensions and increasing currents, the current densities in lower metal layers are also now in the range where EM effects are manifested. EM effects are visible at current densities of about $1\,MA/cm^2$, and such current densities are seen in the internal metal wires of standard cells, resulting in cell-internal signal EM (Jain and Jain 2012). These high current densities arise because local interconnect wires within standard cells typically use low wire widths to ensure compact cell layouts. In short metal wires, such effects were traditionally thought to be offset by Blech length considerations (Blech 1976), but for reasons discussed on Sect. 3.3.3, such effects do not help protect intracell wires in designs at deeply scaled technology nodes. Second, EM has become increasingly important in signal wires, where the direction of current flow is bidirectional. This is due to increased current densities, whose impact on EM is amplified by Joule heating effects (Agarwal et al. 2014; Lee 2012b), since EM depends exponentially on temperature. Therefore, the current that flows through these wires to charge/discharge the output load can be large enough to create significant EM effects over the lifetime of the chip.

For signal nets with bidirectional current flow, the time-average of the current waveform is often close to zero. However, even in cases where the current in both directions is identical, it is observed that EM effects are manifested, as presented in Sect. 3.1.

Intracell power networks are also associated with EM concerns. In going down to deeply scaled technology nodes, the current through the power rails of the cells has remained roughly constant while the cross-sectional area of power rails has decreased, causing the current density in power rails to increase (Wang et al. 2014). Moreover, the power rails are generally subjected to a unidirectional current flow, referred as DC electromigration, which acts more aggressively in causing electromigration (Pelloie 2013).

In the cell library used in this work, we can see high current densities on the Vdd and Vss power rails as well as on signal wires, reducing the lifetime of the cells. For example, we compute signal wires in an INV_X4 (inverter with size 4, where the transistors are four times larger than the minimum transistor size) cell to have an effective average current density of $1.8\,\mathrm{MA/cm^2}$ at 2 GHz, while power wires have an effective current density of $2.15\,\mathrm{MA/cm^2}$ in a 22 nm technology. This switching rate is very realistic, and can be seen in, for example, clock buffers in almost any modern design, as well as in cells that switch at 25 % probability in a 4 GHz design.

While the cell-internal signal EM problem has been described in industry publications such as Jain and Jain (2012), its efficient analysis is an open problem. In this work, we study the problem of systematically analyzing cell-internal signal EM due to both AC EM on signal wires and DC EM on the Vdd and Vss rails of the cells. We devise a solution that facilitates the analysis and optimization of cell-internal signal EM for a standard-cell library based design. We first develop an approach to efficiently characterize cell-internal EM over all output, Vdd, and Vss pin locations within a cell, incorporating Joule heating effects into our analysis. We then formulate the pin optimization problem that chooses cell output pins during place-and-route so as to maximize the design lifetime.

To motivate the problem, we use the example of the INV_X4 (inverter with size 4) cell, whose layout is shown in Fig. 1.5a, from the 45 nm NANGATE Open cell library (Nangate 2011). The input signal A is connected to the polysilicon structure. The layout uses four parallel transistors for the pull-up (poly over p-diffusion in the upper half of the figure) and four for the pull-down (poly over n-diffusion in the lower half of the figure), and the output signal can be tapped along the H-shaped metal net in the center of the cell. The positions where the output pin can be placed are numbered 1 through 7, and the edges of the structure are labeled e_1 through e_6, as shown in the figure. Since the four PMOS transistors are all identical, by symmetry, the currents injected at nodes 1 and 5 are equal; similarly, the NMOS-injected currents at nodes 3 and 7 are equal.

Considering the cell-internal signal EM. When the output pin is at node 4, the charge/discharge current is as shown in Fig. 1.5b. The distribution is similar when the pin is at node 2 or 6 (because of the asymmetry in the parasitics at these two nodes is very small), except that the current direction in e_3 and e_4, respectively, is in opposite direction (reversed). Moving the pin changes the current distribution in e_1–e_6. The currents and the TTF are calculated for each edge, i.e., each wire segment is analyzed separately. If the pin is at node 3 (Fig. 1.5c), since the rise and fall discharge currents have similar values, the charging current in edge e_2 is about $2\times$ larger than the earlier case, while the discharging current is about the same (with opposite direction).

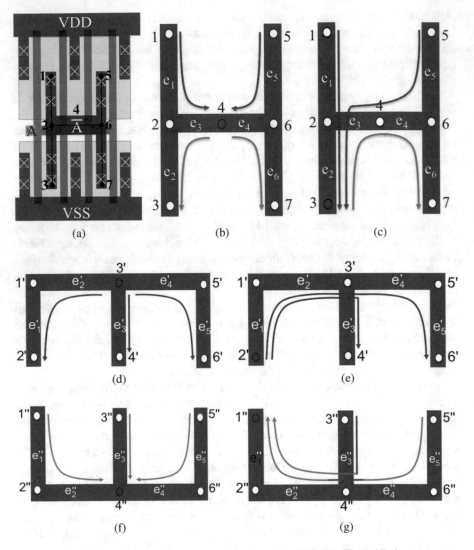

Fig. 1.5 (**a**) The layout and output pin position options for INV_X4. Charge/discharge currents when the output pin is at (**b**) node 4 and (**c**) node 3. The *red* [*blue*] *lines* represent rise [fall] currents. (**d**) The Vdd pin position options for INV_X4 and the currents when the Vdd pin is at node $3'$ and (**e**) node $2'$. (**f**) The Vss pin position options for INV_X4 and the currents when the Vss pin is at node $4''$ and (**g**) node $1''$

As quantified in Chap. 3, the larger peak current leads to a stronger net electron wind that causes EM, resulting in a larger *effective average current*, and therefore, a lower lifetime. In fact, based on exact parasitic extraction of the layout, fed to SPICE (thus including short-circuit and leakage currents), the average effective EM current through e_2 is 2.43× for 22 nm technology larger than when the pin is at node 4. Accounting for Joule heating, this results in a 2.79× lifetime reduction. For the Vdd and Vss pins, a similar effect occurs when the pin position is changed.

Next, we consider EM on the supply wires. Figure 1.5d, e represents the Vdd rail, where the Vdd pin can be placed on the nodes numbered 1' through 6'. Figure 1.5d shows how the charge current is flowing through the edges when the Vdd pin is placed at node 3'. We can see that the current flows are symmetric for this pin position. Since the edge e'_3 supplies two transistors, as shown in Fig. 1.5a, the current flowing through e'_3 is larger than the current flowing through the other edges, which each supply just one transistor. Thus, the edge e'_3 is the critical edge when the Vdd pin is placed at node 3'. Figure 1.5e shows the current flowing through the edges when the Vdd pin is placed at node 2'. In this case, the current flowing through edge e'_1 supplies three of the four transistors, is 3× larger than the current flowing through this same edge when the pin is at node 3'. Thus, this is the critical edge for this pin position, reducing the lifetime of the cell by 2× compared with the lifetime when the pin is placed at node 3'.

Similarly, the Vss rail of the INV_X4 cell is represented in Fig. 1.5f, g. The Vss pin can be placed on the numbered nodes 1'' through 6'', and the currents being discharged through the edges by the Vss pin placed at node 4'' are shown in Fig. 1.5f. Using a similar argument as for the Vdd case, moving the pin from node 4'' in Fig. 1.5f to pin 1'' in Fig. 1.5g changes the critical edge from e''_3 to e''_1, and the lifetime again degrades by about 2×.

So, the contributions of this work can be summarized as follows:

1. A study of the problem of analyzing the EM effects inside standard cells

   - for signal wires, output wire, where the current that flows through the local interconnect wires to charge/discharge the output load can be large enough to create significant EM effects over the lifetime of the chip, and;
   - for the supply wires, Vdd and Vss, where their lifetime is dependent on the pin position.

2. The EM modeling incorporating Joule heating effects and the current divergence to estimate the lifetime of the signal and supply wires (Chap. 3).
3. An approach to efficiently characterize cell-internal EM over all output, Vdd, and Vss pin locations within a cell using a reference pin position. Wherein a graph-based algorithm is used to compute the currents through each edge when the pin position is moved from the reference case to another location (Chap. 4).
4. The pin placement optimization problem formulation (Chap. 5)

   - whereby place-and-route chooses cell signal output pins in such a way that the lifetime of the overall design is maximized;
   - for the supply pins, Vdd and Vss, where the pins are placed with the objective to optimize the circuit lifetime.

The pin positions to be avoided by the router will depend on the logic of the gate and on the wire shape. The pin positions that produce a high current density in one or more wire segments will be avoided. The critical pin positions are avoided just changing the Library Exchange Format (LEF) file of the cells, the cell layouts are not changed. In this way, just the routing step of the design flow changes to consider the TTF-optimized LEF file.

5. Cell layout optimization changes are suggested to improve the cell robustness from the EM perspective. These robust cells can replace the critical cells affected by EM in the circuits, if the circuit should have a larger TTF than the TTF achieved just by the pin position optimization. For this, the current flows and the EM effects behavior for different pin positions and different logic gate are presented (Chap. 6).

6. At circuit level, we present an analysis of the EM effects on different metal layers and different wire lengths for signal wires (nets) that connect cells. The delay behavior and how the average current reduces through the wire are also reported in this analysis (Chap. 7).

1.4   Monograph Outline

This monograph is organized as follows:

A review of existent works that address the EM problem is presented in Chap. 2. Some concepts of the digital circuit physical design flow and the steps where techniques to mitigate the EM effects could be applied are presented in Sect. 2.1.

Chapter 3 describes how the cell-internal EM is modeled in our work incorporating Joule heating effects and the current divergence.

The current calculation approach for our cell-internal EM analysis flow is presented in Chap. 4, where the algebras to calculate the average and RMS currents for different pin positions are described. Our graph-based algorithm that computes the current through each edge when the pin is moved is also presented in this chapter. Next, a method for optimizing the circuit lifetime using incremental layout modifications is proposed. The circuit lifetime can be increased by placing the output, Vdd, and Vss pins appropriately, avoiding the critical pin positions that reduce the lifetime of the cells by EM.

The implementation flow is then discussed in Chap. 5.

The experimental results of our approach are shown in Chap. 6. We are presenting an EM analysis at circuit level for the nets that connect the cells.

The tests are considering different metal layers and different wire lengths for the nets (Chap. 7).

At last, conclusions and future works are presented in Chap. 8.

Chapter 2
State of the Art

EM is a well-known problem and many methods have been proposed to model and to mitigate the EM effects in different design stages, as Sect. 2.1 presents, for different types of interconnections as presented in Sect. 2.2.

2.1 Mitigating the EM Effects in Different IC Design Flow Stages

There are two main methodologies used in IC design to develop application specific integrated circuits (ASICs): the full-custom and the standard cells (Rabaey et al. 2002; Weste and Harris 2005).

In a full-custom methodology, each individual transistor and interconnections in the circuit is designed in a customized way. Every design stage is carefully done to obtain the best possible circuit option in terms of area, speed, and power consumption. Transistors are deliberately laid out in chips most compactly, spending months by many designers and engineers (Chen 1999), increasing significantly the final design cost. The full-custom methodology is generally used to construct the standard-cell library and in analog designs.

Traditional standard-cell-based synthesis flow has been used in the industry and academia for a very long time. The cells are taken from a pre-designed and pre-characterized cell library, generally, designed-by-hand. Standard-cell based synthesis flows are known to be very reliable and predictable.

The standard-cell based synthesis flow generally follows the methodology presented in Fig. 2.1 and can be divided into two main stages: logic synthesis and physical synthesis. The logic synthesis receives as input a circuit description (in VHDL, Verilog), the design constraints, and the cell library file (.lib, .db) and generates a structural description composed by logic gates and registers. This step explores the selection of logic gates from a cell library to represent the best result

© Springer International Publishing AG 2017
G. Posser et al., *Electromigration Inside Logic Cells*,
DOI 10.1007/978-3-319-48899-8_2

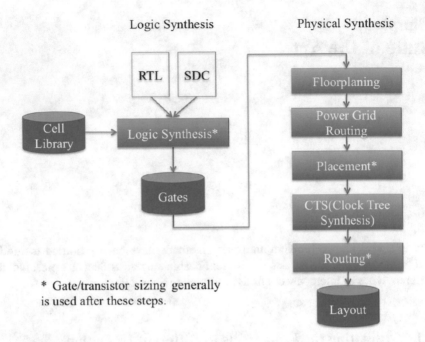

Fig. 2.1 Standard-cell based synthesis flow (Rabaey et al. 2002; Weste and Harris 2005)

considering the design constraints. The physical synthesis generates the circuit layout from the structural description provided by the logic synthesis following these steps (Kahng 2011):

- Floorplanning: usually is the first step of the physical synthesis. It is responsible to define the position of the high level blocks in the total circuit area. The I/O pads position and power supply network are also defined in this step;
- Power grid routing: determines not only the layout of the power-ground distribution network, but also the placement of supply I/O pads or bumps (Kahng 2011);
- placement: is responsible to place the cells (gates) that compose the circuit in the circuit area aiming to reduce the wire length;
- CTS—Clock Tree Synthesis: performs the clock tree routing in synchronous circuits to distribute the clock signal throughout the circuit;
- Signal routing: performs the signal routing among the cells respecting the connections description from the circuit netlist;
- Gate/transistor sizing: is used throughout the design flow to correct timing errors and to optimize the design. It is commonly used after logic synthesis, placement, and routing (Lee 2012a. The gate/transistor sizing has as objective to determine the best size for each logic gate (transistor) of the circuit considering the current that must be supplied to charge the attached load (Flach et al. 2013 2014; Posser et al. 2012 2014a; Reimann et al. 2013).

After the routing, the layout of the circuit is complete. However, the characteristics of the circuit have to be verified to know if the circuit will operate as required. For this, verifications are used to check if the circuit meets the specifications generated early. If the product requirements are not met, a rework is required (Butzen 2012; Kahng 2011).

EM-aware optimization is an important part of high reliability circuit design (Xie et al. 2012), where technology nodes smaller than 65 nm require power integrity (e.g., IR drop-aware timing, electromigration reliability) analysis flows (Kahng 2011). Thus, we have to include some steps in the IC design flow to consider and reduce the EM problem in the circuits. In this work, the circuit lifetime is improved under cell-internal EM in the routing step of the standard-cell-based synthesis flow. Chapter 5 shows in details the implementation flow used in this work. Previous works in the literature have applied techniques through the IC design flow to mitigate the EM.

Balhiser et al. (2005) present a method and system for enabling efficient identification of nets that are at risk of failure due to the effects of electromigration permitting targeted assessment and redesign of the identified nets before the final steps of the design.

An early stage calculation to determine the worst-case bounds on interconnect segment currents under the effect of EM is presented in Jerke and Lienig (2010). This early stage calculation enables nets to be separated into critical and non-critical sets. Only the set of critical nets, which is typically considerably smaller, requires subsequent special consideration during physical design and layout verification due to current density design limits. The algorithms used are fast enough to run on every net and can be used to the pre-layout identification of nets that are potentially troublesome and may need sizing.

Sections below present the most relevant EM related works in details with their techniques in different steps of the design flow aiming to produce EM-aware designs.

2.1.1   *Managing Electromigration in Logic Designs*

Considering the standard-cell-based synthesis flow (Rabaey et al. 2002; Weste and Harris 2005) for integrated circuits design, Barwin and Bickford (2013) present an invention that provides a method to avoid potential EM violations during an early design step, before the logic synthesis. Available circuit information is used to modify maximum capacitance or maximum output slew rate that each individual cell is allowed to drive, it is possible to design a circuit early in the design cycle to avoid EM violations. And, by knowing EM violations prior to arriving at the layout of the circuit design, a considerable time and expense during later design stages can be saved, e.g., simulation and testing, by not having to redesign the circuit.

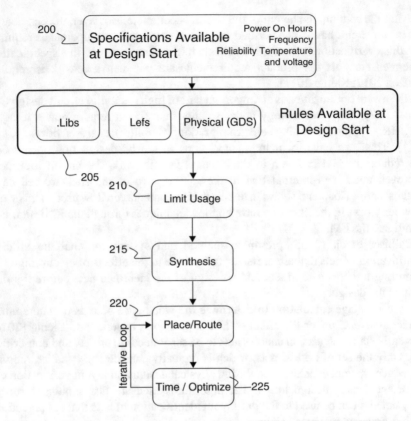

Fig. 2.2 Flow diagram implementing processes proposed (Barwin and Bickford 2013)

On the other hand, this methodology can super estimate the EM violations making the circuit works below its maximum performance. Or, if the EM violations are super estimated, an additional step to evaluate the EM effects on the final circuit layout has to be included.

The proposed flow is presented in Fig. 2.2, where the design specifications and rules are set at blocks 200 and 205. At block 210 is the contribution of the work, the program control will limit usage, e.g., limit the output capacitance that the cell can drive to avoid EM violations, by calculating the maximum output slew or output capacitance on each output pin using the available specifications and rules. At block 215, the logic synthesis is performed using the constraints from block 210. At block 220, the place and route steps are executed. At block 225, the program control will time and optimize the circuit. Blocks 220 and 225 will undergo an iterative loop until the design meets its specifications.

2.1.2 Electromigration Impact in Future Technologies

The EM problem during layout synthesis is addressed in Lienig (2013), focusing on basic design issues that affect electromigration during interconnect physical design. The aim is to increase current density limits in the interconnect by utilizing electromigration-inhibiting measures, such as short-length and reservoir effects (Sect. 2.2.4). Exploitation of these effects at the layout stage provides partial relief of EM concerns in today's design flows. Design tools can significantly improve the EM robustness of the generated layout by utilizing EM-optimized layout configurations as constraints during synthesis steps, such as routing. It is believed that the inclusion of EM-specific requirements in the physical design can provide relief from severe reliability constraints in future technologies.

Lienig (2013) presents flow options to analyze the impact of electromigration on circuit reliability as Fig. 2.3c shows. These options are used in different synthesis steps of the digital design flow (Figs. 2.1 and 2.3a). The verification steps presented in Fig. 2.3b ensure that the circuit acquires the required electrical characteristics and functions, and meets the reliability and manufacturability criteria. Lienig (2013) cites different ways to do an EM analysis, but specific techniques are not presented to mitigate the EM effects. Moreover, the cell-internal EM is not referred.

The flow options to analyze the impact of the EM have only been partly supported to date by layout tools. However, "Sign-off DRC w/ EM-rules" and "Sign-off SPICE Simulation" with subsequent current density verification are now standard functions in the state-of-the-art digital layout tools. For example, Synopsys IC Compiler offers a signal electromigration analysis and repair to improve design reliability (Synopsys 2014a) while Cadence Encounter (Summers 2013) now it's called Innovus, when running power and rail analysis take into account the electromigration (EM) model file where the various metal width and via sizes are considered generating an

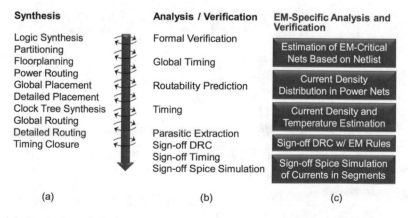

Fig. 2.3 Synthesis-analysis loops in the design flow for digital circuits. The critical steps—physical synthesis (**a**) and analysis (**b**)—are shown, supplemented by options to address current density and other electromigration issues (**c**) (Lienig 2013)

EM analysis for that design. These functions are also available as stand-alone verification tools (for example, Apache Totem MMX, Mentor Calibre PERC, and Tanner HiPerVerify).

2.1.3 Smart Non-default Routing for Clock Power Reduction

Kahng et al. (2013b) study the EM problem in clock networks. For new technologies the non-default routing rules (NDRs) are integrated to clock network synthesis methodologies. NDRs apply wider wire widths and spacing to address electromigration constraints, and to reduce parasitic and delay variations. Kahng et al. (2013b) propose a "smart" NDRs (SNDR) to substitute narrower-width NDRs for selected clock segments while maintaining all skew, slew, insertion delay, and EM reliability criteria. A wire sizing problem is formulated of choosing per-segment NDRs to minimize the capacitance of the clock tree subject to electrical and reliability constraints, and a solution to this problem for each subnet is given. Then, the SNDR approach is extended to the entire clock tree by propagating skew constraints from downstream to upstream subnets. Also, a methodology to apply SNDRs without disruption of standard clock network synthesis flows is used.

The flow that implements SNDRs proposed in Kahng et al. (2013b) uses a commercial place-and-route tool and is presented in Fig. 2.4. Clock tree synthesis with a given fixed NDR is performed for an input design. The locations of buffers and flip-flops are extracted after routing, which are the sources and sinks of clock nets, along with all clock wire segments, from the routed DEF (SI2 2009) of

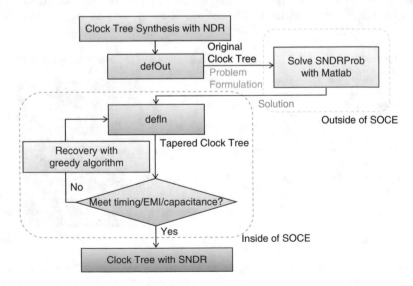

Fig. 2.4 Overall implementation flow (Kahng et al. 2013b)

the implemented design. From this information, the optimization instances are constructed with appropriate timing (delay, skew, and slew) and EM constraints. Solving each optimization instance to obtain wire sizing solutions for every edge of each given subnet. Then the original DEF file with the obtained wire sizing solutions for each subnet is updated. The modified DEF is read back into the P&R tool, and RC values are recalculated. With the updated RC values, electrical/timing or EM reliability violations are check; if there are any violations caused by an SNDR on a wire segment, that segment is reverted to its original NDR.

SNDR methodology was executed with a 32/28 nm library (Synopsys 2014b) and open-source benchmark netlists. SNDR presented power and capacitance reduction versus the fixed-NDR (4W5S) methodology used in a commercial tool.

2.1.4 Impacts of Electromigration Awareness

Kahng et al. (2013a) present two studies: EM lifetime versus performance with fixed resource budget and EM lifetime versus resource with fixed performance. AES, DMA, and JPEG designs are used as example for TSMC 45GS and 65GPLUS technology libraries. Kahng et al. (2013a) show that the performance scaling achieved by reducing the EM lifetime requirement depends on the EM slack in the circuit, which in turn depends on factors such as timing constraints, length of critical paths, and the mix of cell sizes. Depending on these factors, the performance gain can range from 10 to 80 % when the lifetime requirement is reduced from 10 years to 1 year, as Fig. 2.5 shows.

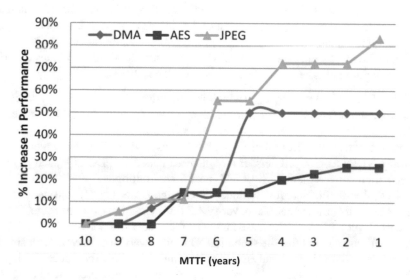

Fig. 2.5 Percentage increase in F_{max} at 45 nm due to reduced MTTF requirement (Kahng et al. 2013a)

Figure 2.7 shows the flow presented in Kahng et al. (2013a) to find the highest possible maximum frequency of a design for a given reduced MTTF requirement. Synopsys DesignCompiler (Synopsys 2013b) and Cadence SOC Encounter (Cadence 2013) flows are used to synthesize the circuits and the thermal analysis is done using Hotspot. The flow used follows the same main steps presented in Figure 2.1. The steps changed to consider the EM are described in details below.

- Characterize current density limits in each metal layer in the technology LEFs based on upper bound on the peak temperature of the design (T_{UB}), switching activity, and reduced MTTF requirements (MTTF$_{red}$) using Black's Equation described in Chap. 3 (Eq. (3.1)).
- Place and route the post-synthesis netlist using the newly characterized technology LEF from the previous step.
- Perform post-route extraction, timing, and signal integrity (SI) analysis. Fix EM violations using P&R tool commands.
- Calculate peak temperature of the design (T_{peak}) from chip and core areas, ambient temperature (T_{amb}), and power using Hotspot (Skadron et al. 2003) calibrated to a 45 nm Qualcomm SoC package.
- Check that all constraints are met.
- If all constraints are met, then increase frequency (decrease clock period in SDC) using binary search. If any constraint is violated, decrease frequency (increase clock period) using binary search.
- If the next frequency has already given a valid solution, then exit, else repeat the flow using a frequency obtained from the previous step.

Kahng et al. (2013a) also study how conventional EM fixes using per net NDR routing, downsizing of drivers, and fanout reduction affect performance at reduced lifetime requirements. This study indicates, e.g., that NDR routing can increase performance by up to 5 % but at the cost of 2 % increase in area at a reduced 7-year lifetime requirement.

Figure 2.6 shows the flow to create NDRs per net; the following steps describe implementation in Cadence SOC Encounter (Cadence 2013).

- Group EM critical nets by I_{rms} (which is a function of the switching activity on the net) and create NDRs for wire width and spacing depending on the extent of I_{rms}.
- Find the extent of current density violations vis-a-vis LEF in each metal layer. Apply NDRs to grouped nets. If any of these nets are "FIXED" (such as clock nets), then convert them to SPECIAL nets for re-routing.
- Select all violating nets for Engineering Change Order (ECO) routing. Perform ECO routing of these nets. Then, verify AC limit violations after re-route. If there are AC limit violations, decrease fanout of these nets and perform ECO routing.
- If violations remain after the second ECO route, swap large drivers with smaller drivers and redo timing analysis. If no timing violations, then accept this new frequency and exit, else reduce frequency and run the flow presented in Fig. 2.7. If frequency $\leq F_{max}$ obtained from the flow in Fig. 2.7, then exit.

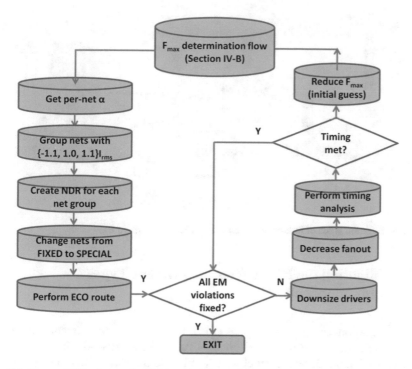

Fig. 2.6 Per-net NDRs flow to fix EM I_{rms} violations (Kahng et al. 2013a)

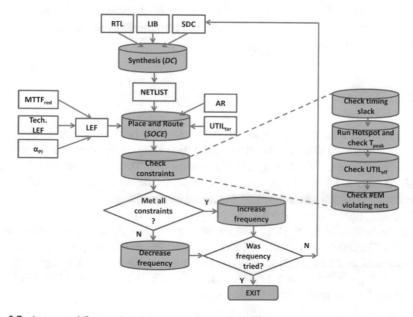

Fig. 2.7 Automated flow to determine F_{max} (Kahng et al. 2013a).

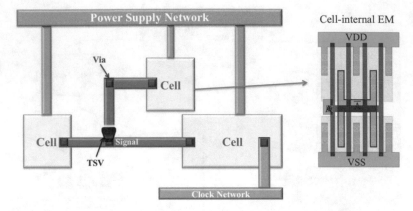

Fig. 2.8 Different types of interconnections within the circuit where EM occurs

2.2 Mitigating the EM Effects in Different Types of Interconnections

Electromigration (EM) is an aging effect taking place in interconnect wires, contacts, and vias in an integrated circuit (Tu 2003). Most works in the literature are considering the different net classes to mitigate the EM effects: TSVs (present in 3D circuits), power supply network, clock network, and vias, as Fig. 2.8 presents, but few works told about the EM in the signal interconnects within a standard cell (*internal-cell EM*) that is the focus of this work. The EM effects for the different net classes and the two works that cite something about the cell-internal EM are related in the next subsections.

2.2.1 TSVs

In the last years several strategies were proposed to mitigate the EM impact on through-silicon-via (TSVs). An electromigration (EM) reliability study for TSV-based 3D ICs and 3D power delivery networks (PDNs) is presented in Zhao et al. (2013). A transient power integrity analysis flow for lifetime prediction is developed, which integrates the EM modeling approach.

Pak et al. (2013) model the EM on TSVs and local vias used together for vertical power delivery. Cheng et al. (2013) propose a framework at architecture level to alleviate EM effects of defective TSVs. At first, the relationship between various TSV defects and EM induced TSV mean time to failure (MTTF) degradation is analyzed. Then, a framework to protect defective TSVs and improve EM MTTF by balancing their current flow directions is proposed.

2.2.2   Power Delivery Network

Due to shrinking wire widths and increasing current densities, there is little or no margin left between the predicted EM stress and that allowed by the EM design rules in the power delivery networks (PDNs) (Fawaz et al. 2013). Most works in the literature are concerned with the EM effects in the power delivery network.

Abella et al. (2008) presented refueling that is a microarchitectural technique to make up for EM in power/ground grids using the self-healing effect that can undo EM. This effect occurs when current flows in both directions of a wire. The parts of the wire that are prone to form voids when current flows in one direction are prone to form hillocks when it flows in the opposite direction. Thus, if the same amount of current flows in each direction, EM self-heals. It consists of injecting current whenever the amount of current in one direction is higher than the amount in the opposite direction.

Other solution based on the electromigration AC healing effect is presented in Xie et al. (2012) to extend the lifetime of power supply networks. An algorithm for the EM-aware design that can be integrated into the standard-cell place and route flow is also presented. The topology of power networks to produce balanced bidirectional current on power rails is changed, as Fig. 2.9 shows, with compensation strip insertion. There are two operation modes: the normal mode and the compensation mode. The current flow directions on power rails are shown in Fig. 2.9a. Both modes are driven by the same set of PADs to prevent PAD number increase. In the normal mode, power is supplied to the block from the Power/Ground (P/G) ring. The transistors connecting PADs and the compensation strip are off, thus the strip is in high-impedance state. In the compensation mode, the PAD supplies the compensation strip, and the P/G ring is in high-impedance state. If a block is too big to meet the IR drop requirement, it can be divided into regular or irregular sub-blocks with their compensation strips connected together as illustrated in Fig. 2.9b. The sub-blocks switch into the normal or the compensation mode simultaneously.

For power grids, the classical approaches have some limitations. The classical approaches do not accurately capture the physics of EM degradation in copper dual-damascene (CuDD) metallization. Moreover, the classical approaches fail to model the inherent resilience in some circuits that may keep functioning even after the failure of some of the wires. Mishra and Sapatnekar (2013) overcome the limitations and drawbacks that the classical circuit-level EM models. For a single wire, their probabilistic analysis encapsulates known realities about CuDD wires, e.g., that some regions of these wires are more susceptible to EM than others, and that void formation/growth show statistical behavior. They apply these ideas to the analysis of on-chip power grids and demonstrate the inherent robustness of these grids that maintains supply integrity under some EM failures.

A mesh model to estimate the MTTF and reliability of power grid under the influence of EM taking into account the redundancies due to its mesh structure is proposed in Chatterjee et al. (2013). To implement the mesh model, a framework to estimate the change in statistics of an interconnect as its effective-EM current

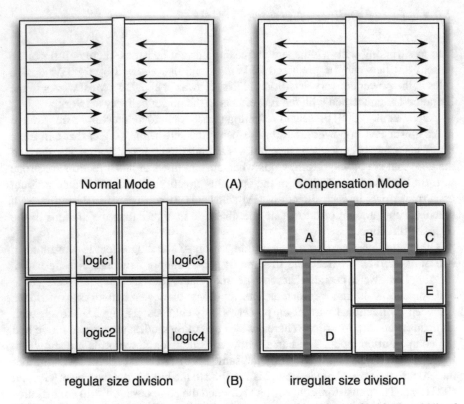

Normal Mode   (A)   Compensation Mode

logic1   logic3

logic2   logic4

regular size division   (B)   irregular size division

Fig. 2.9 (a) A vertical Power/Ground (P/G) strip (compensation strip) is added in the middle of the layout with two working modes (normal mode: power is supplied to the block from the P/G ring with the compensation strip in high-impedance state; compensation mode: the PAD supplies the compensation strip, with the regular P/G ring in high-impedance state); (b) chip layout divided into regular or irregular sizes with power grid (Xie et al. 2012)

varies was developed. Moreover, a fast and exact approach to update the voltage drops is presented. The proposed mesh model accounts for the redundancies in the grid based, that gives a longer lifetime than a series system, and thus obtains a more realistic estimate of grid's lifetime and reliability.

Another EM checking approach that reduces the pessimism in EM prediction for power/ground grids is presented in Fawaz et al. (2013). A framework for EM checking is proposed, allowing users to specify conditions-of-use type constraints that help capture realistic chip workload and which includes the use of a novel mesh model for EM prediction in the grid, instead of the traditional series model employed for EM reliability estimation. To compare the two models, the series system TTF was computed at the optimal points obtained, and the mesh model TTF was 2–3× larger than that of the series system.

A temperature-and variation-aware electromigration analysis (T-VEMA) tool for power grid wires is presented in Li et al. (2015b). T-VEMA performs a two-stage interconnect thermal analysis and performs an electromigration (EM) lifetime calculation on ideally manufactured mortal wires on the basis of thermal effects. At the end, tool analyzes process variation effects on EM reliability at global and local levels and reports variation tolerances of EM-sensitive power grid wires.

Mishra and Sapatnekar (2016) address the issue of "EM mortality" in interconnects. A wire may be prevented from being mortal under EM if the maximum stress build-up, corresponding to the equilibrium between the current-induced forward stress and the back stress due to the gradient in atomic concentration along the wire, does not exceed the critical stress due to void nucleation. Alternatively, it may also not be mortal if the stress build-up does not exceed the critical stress over the lifetime of the circuit. A new efficient approach, based on multiple filters, is developed for determining the mortality of wires in a circuit. These filters greatly reduce circuit analysis time by predicting which wires can never be mortal over the circuit lifetime under its operating conditions so that detailed analysis must only be performed over a small subset of all interconnects.

2.2.3 Clock Network

At advanced process nodes, NDRs are integral to clock network synthesis methodologies. One of the reasons for this is that the signal electromigration (EM) limits are violated by minimum-width wires when large buffers (e.g., 32×) are used to drive large fanouts (e.g., anywhere from 16 to 40 loads for each clock buffer instance in a typical buffered clock tree solution). To satisfy EM limits, wider wiring must be used (Kahng et al. 2013b). In this way, NDRs apply wider wire widths and spacing to address electromigration constraints, and to reduce parasitic and delay variations. However, wider wires result in larger driven capacitance and dynamic power. A practical methodology to apply "smart" NDR in standard clock tree synthesis flows is formulated in Kahng et al. (2013b) to minimize wire capacitance under skew, slew, delay, and EM constraints. The problem is solved for each subnet of a given clock tree and the potential for capacitance and power reduction is quantified. More details of this work are presented in Sect. 2.1.3.

An early work that also is concerned with EM in clock networks is presented in Pullela et al. (1995), where an EM constraint is considered in their low-power clock tree design methodology. They optimize a clock tree with buffer insertion by decreasing wire width while satisfying bounds on process variation-dependent skew and current densities that affect directly the EM.

2.2.4 Vias

Particular attention needs to be paid to vias and contact holes, because generally the ampacity[1] of a (tungsten) via is less than that of a metal wire of the same width. Moreover, migration velocities in the via material, the diffusion barrier, and the metal wire differ. Hence, vias are one of the prime points of void nucleation and thus EM failures (Lienig 2013). To compensate for this increased vulnerability and to address typical manufacturing and yield issues, double or multiple vias are often deployed.

The situation can be improved when redundant vias are used (Li et al. 2005; Lienig 2006). The via array geometry is crucial: multiple vias need to be arranged so that the resulting current flow and thus EM degradation is distributed as evenly as possible throughout the parallel vias (Lienig 2006). However, for the advanced technologies, the minimum required via to via spacing does not always scale with the via size and line width and the metal lines may not be wide enough to accommodate multiple vias along the line width.

One of the solutions for this problem is to offer multiple via size options in addition to the regular square via, like large square via and rectangular bar via (Li et al. 2014) as shown in Fig. 2.10. While the large and bar via options are very helpful and efficient for circuit designs with better reliability and scaling, they also

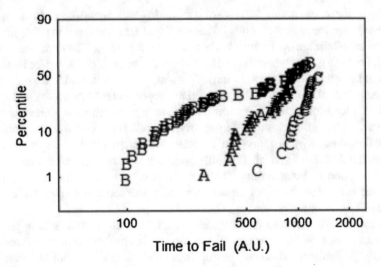

Fig. 2.10 EM performance variation with different via sizes. *A*—two regular square vias along line width; *B*—one rectangular bar via with 2× the cross-sectional area of a regular square via; *C*—one larger square via with 4× the cross-sectional area of a regular square via (Li et al. 2014)

---

[1]Ampacity is the maximum amount of electrical current a conductor or device can carry before sustaining immediate or progressive deterioration.

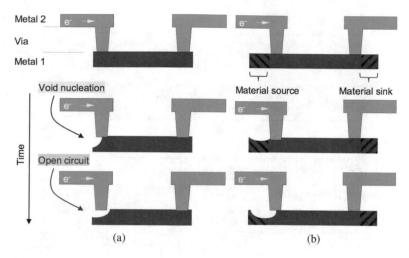

Fig. 2.11 The reservoir effect in (**b**) extends the lifetime of the via configuration compared to a regular via without layer overlaps (**a**) (Lienig 2013)

bring in challenges and complexities to the manufacturing process control. Instead of optimizing the integration process window to single size via, a balance for all via sizes is needed.

Other solution to improve the lifetime of vias is presented in Nguyen et al. (2002) where the "reservoir effect" is used to improve the lifetime of multiple-level interconnects. Figure 2.11 shows how reservoir effect can be used considering the direction of current flow. Reservoirs can act as sources that have to be drained before voids generated by electromigration become critical to the circuit. This does not actually decrease the effects of electromigration itself, but prolongs the time to failure due to void growth (Lienig 2013).

The reservoir effect is also leveraged by using multiple vias, as we presented before from Li et al. (2005) and Lienig (2006) works, since via arrays have a larger reservoir area than a single via. Experimental results in Nguyen et al. (2002) indicate that the prolonged lifetime achieved is more a result of the increased reservoir area than of current sharing between vias.

The generation of voids in the vicinity of vias is strongly influenced by the geometry of the via configuration. Depending on whether a specific wire segment is connected from above or below, the configuration is called via-above or via-below, respectively (Choi et al. 2004; Thompson 2008). Figure 2.12 shows that even a low-volume void causes a failure if placed directly underneath the via, whereas a void in a via-below configuration has to grow larger before the interconnect is destroyed (Lienig 2013).

Park et al. (2010) present an EM check method named Via Node Vector that addresses the EM interactions and checks the EM reliability at the lead connection

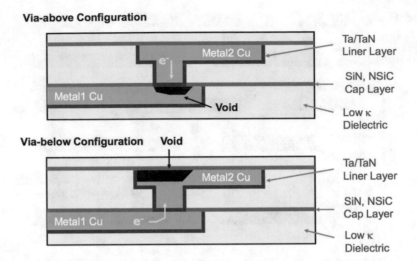

Fig. 2.12 Via-above and via-below configurations with their different damage locations partly due to the interface/surface diffusion prevalent in copper wires (Lienig 2013)

sites (called via node). Moreover, three new factors are introduced: length (FL), width (FW), and interaction (FB). It is a model that considers the divergence of the current flowing in the interconnects taking into account the vias.

2.2.5 Signal Interconnects

AC electromigration in signal wires is gaining more attention in the latest years as for new technologies EM on signal wires is becoming more and more a big concern (Zhou et al. 2015). Below are some of this important studies in AC EM.

Mishra and Sapatnekar (2015) address signal interconnects that carry bidirectional (AC) currents as they transport logic signals within the digital system. For these wires, it is shown that EM is not only a catastrophic failure problem, but also is capable of causing parametric shifts in circuit performance over time. HSPICE-based Monte Carlo simulations are performed on a standard on-chip structure to quantify the impact of EM on circuit performance degradation. Although the damage due to EM degradation under bidirectional currents is reduced relative to the unidirectional current case due to partial EM recovery, it is demonstrated that, depending on the level of recovery, the circuit performance may degrade beyond acceptable limits and can be comparable to other transistor degradation mechanisms.

A method of characterizing an electromigration (EM) parameter to be used in an integrated circuit (IC) design, including inputting a layout of a wire layer and identifying a signal including electrically parallel paths, connected to an output of the signal gate is presented in Barwin et al. (2015). A maximum possible current for each of the paths is determined based on widths for each of the paths. The path that is most limited in its current carrying capacity by possible EM failure mechanisms is identified, then a possible maximum current output to the identified limiting path is stored in a design library as the EM parameter.

White et al. (2016) disclosed methods, systems, and articles of manufacture for implementing electronic circuit designs with electromigration awareness for power and signal interconnects. Some embodiments perform schematic level simulations to determine electrical characteristics, identify physical parasitics of a layout component, determine the electrical or physical characteristics associated with EM analysis on the component, and determine whether the component meets EM related constraints while implementing the physical design of the electronic circuit. Further adjustment(s) can be determined to the component or related data where the EM related constraints are not met and/or present the adjustment in the form of hints.

2.2.6 Cell-Internal EM

To consider the EM inside of the cells is different than to consider in other interconnections because the current flowing through the wire segments depends on the logic of the gates.

A retargetable methodology for IP-internal EM verification is presented in Jain et al. (2016), where the authors show how for a given cell, the current (AVG + RMS) flow can be computed through the resistors under arbitrary switching scenarios, for any load/slew combination, and under effect of parasitic RC loading. Generic switching rates for various pins of the IP are comprehended, including aspects of clock gating. By incorporating the impact of arbitrary parasitic loading and an intelligent way of coming up with the effective pin capacitance of load cells a high accuracy with respect to SPICE was achieved. The methodology was shown to be flexible, in terms of allowing on-the-fly retargeting for the reliability. The complete data generation process at library level is expedited by application of cell-response modeling. Results on a 28-nm production setup were shared, to demonstrate significant relaxation in terms of violations, along with close correlation to SPICE.

For standard cells, we have more two works talking about cell-internal EM Jain and Jain (2012) who cites that it is important to consider the EM effects inside of the cells, but it is not presented a solution for that. Domae and Ueda (2001) saw the EM effects in the wires inside of the cells when an inverter was fabricated and tested, and the suggested some layout changes to reduce the EM effects. These works are detailed as follows.

2.2.6.1   Accurate Current Estimation for Interconnect Reliability
Analysis (Jain and Jain 2012)

The section "standard-cell EM analysis and modeling" in Jain and Jain (2012) cites that robust design of standard cells itself is at the crux of an SoC's design integrity. So, the standard cells need to be reliably operating at the guaranteed conditions. During the process of library design, utmost care needs to be taken to set the correct EM targets for different cells. These targets could be in terms of the maximum operating load, frequency, voltage, input–output slew, temperature, and lifetimes.

To enable performing the standard-cell-internal EM checks at chip level a method of modeling the safe frequency of the cell at different operating points is used. This information can be used to identify EM-critical cell instances and take corresponding design fixes. Good timing constraints, like a maximum frequency, are very effective in limiting the EM-critical cells, which are usually instances operating at high loads, frequencies, or bad slews (typically less than few hundreds).

Jain and Jain (2012) cite that the available models to estimate RMS and AVG current are not accurate to predict the EM safety of standard cells and suggest to use SPICE simulation. As SPICE simulation is very accurate and can also be runtime efficient if the initial filtering of non-EM-critical cells is done properly, through library characterized data.

2.2.6.2   CMOS Inverter and Standard Cell the Same
(Domae and Ueda 2001)

Domae and Ueda (2001) present a patent that tested the reliability of a CMOS inverter with the structure shown in Fig. 2.13. As a result, an interconnection failure was spotted in the interconnection portion of the CMOS inverter. It was also found that this interconnection failure in the CMOS inverter on the last stage in a standard cell composed of a plurality of CMOS inverters. The void was formed in an interconnection region near a contact connected to the source or drain of the p- and n-channel MOS transistors $TR1$ and $TR2$ as shown in Fig. 2.13. Specifically, large voids were formed in the power line 101 near the first contact 102 connected to the source of the p-channel MOS transistor $TR1$ and in the output signal line 105 near the fourth contact 109 connected to the drain of the n-channel MOS transistor $TR2$.

In Fig. 2.13 the broken line arrow indicates the electrons flow direction and the line arrow indicates the current direction, i.e., electron flow and current are in the opposite direction. The current flows bidirectionally between a branch point 105b and the output terminal 105a in the output signal line 105, but only unidirectionally between the third contact 106 and the branch point 105b and between the fourth contact 107 and the branch point 105b. Accordingly, electromigration might possibly happen in these regions of the output signal line 105, even though these regions are short.

Fig. 2.13 A plan view of a
CMOS inverter layout
(Domae and Ueda 2001)

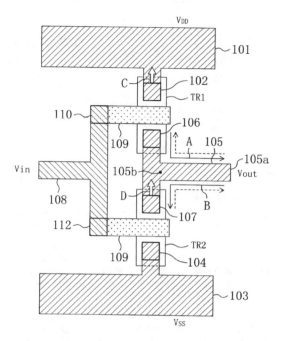

In addition, in the downsized interconnection structure for a CMOS inverter, the line width of an interconnect is substantially equal to the width of a contact as shown in Fig. 2.13, contacts 102 and 104. Comparing to the conventional interconnection structure where the contacts have a metal enclosure, a void is less likely to expand in a portion of a metal interconnect, from which electrons flow out through the end of a contact. This is because while the metal atoms are moving in response to the momentum of electrons, the vacancies left by the metal atoms are filled in with those supplied from the regions of the metal interconnect surrounding the contact. In contrast, in the interconnection structure shown in Fig. 2.13, a void is more likely to expand in that portion of the metal interconnect. This is because while the metal atoms are moving in response to the momentum of the electrons, no metal atoms are supplied from those regions of the metal interconnect surrounding the contact.

Figure 2.14 shows the amount of current flowing through the power line and output signal line increases proportionally to the operating frequency. However, since the operating frequency is very likely to go on rising, the amount of current flowing through an output signal line will exceed the permissible current. Thus, looking this figure from 2001, it was expected that electromigration on an output signal line would be a serious problem in the near future. And this is what our work is presenting.

A solution to improve the reliability in the output wire and other one for the power and ground lines are presented in this invention. To reduce the electromigration effect in the output wire for the inverter in Fig. 2.13, the second output signal line 105 is connected to the contact 107, close to the drain of the NMOS

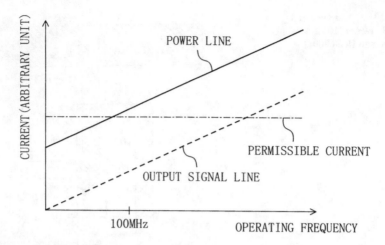

Fig. 2.14 The relationships between the operating frequency of a CMOS inverter and the amount of current flowing through a power line and between the operating frequency and the amount of current flowing through an output signal line (Domae and Ueda 2001)

transistor. So, a current flowing through the output terminal of the inverter into the second output signal line is directly supplied to the contact (107). Thus, electrons and metal atoms, which exist in a portion of the first output signal line near the contact, do not move toward the first output signal line (line from 107 to 106 contact), then electromigration does not happen in that portion of the first output signal line near the contact 107. For the power and ground lines, electromigration happens in a portion of the power (101) and ground (103) lines near the contacts 102 and 104, Fig. 2.13. In such a case, by increasing the line width of that portion of the power/ground line near the contact, it is possible to prevent a void from being formed.

2.3   Summary of Related Works

The main works that have contributed for our work and their characteristics are presented in Table 2.1.

Table is marking the works by their characteristics:

- if works verify and/or analyze EM effects on power networks, signal wires, or cell-internal EM.
- if works are presenting corrected-by-construction solutions.
- if work is presenting fixing strategies for power networks, signal wires, or for cell-internal EM.

Table 2.1 Main related works and their main contributions

Work	Verification/Analysis			Corrected-by-construction	Fix strategies for		
	Power	Signal	Cell-EM		Power	Signal	Cell-EM
Domae and Ueda (2001)			X				X
Balhiser et al. (2005)		X					
Abella et al. (2008)		X		X		X	
Jerke and Lienig (2010)		X		X			
Park et al. (2010)		X				X	
Jain and Jain (2012)			X				X
Xie et al. (2012)	X			X	X		
Mishra and Sapatnekar (2013)	X						
Chatterjee et al. (2013)	X						
Fawaz et al. (2013)	X				X		
Kahng et al. (2013b)		X		X		X	
Barwin et al. (2015)		X				X	
Li et al. (2015b)	X				X		
White et al. (2016)	X	X			X	X	
Mishra and Sapatnekar (2015)		X				X	
Jain et al. (2016)			X				
Cadence (2016)			X				
Synopsys (2016)			X				
This work			X	X			X

As table shows, our work is the unique one that is analyzing, modeling, and presenting a corrected-by-construction solution for cell-internal EM. We are presenting here a method to reduce the cell-internal EM effects by placing the output, Vdd and Vss pins in a cell. We are modeling the EM effects inside of cells and then analyzing the whole circuit to optimize the lifetime by reducing the EM effects.

2.4 Conclusions

In the era of striking for full circuit design automation, it is important to incorporate EM aware circuit design strategies in the design, checking, and verification tools. It is imperative for the electronic design automation (EDA) tools to be able to recognize and identify those circuits with critical EM implementations. In reality, this task is so complicated; none of the existing EDA tools can perform and optimize

EM analysis on the whole circuit, considering all the cell layouts and connections. To take more advantage of different layouts with different EM benefits, it often needs human intervention to avoid having too aggressive designs or keeping too much EM margins by the EDA tools (Li et al. 2014).

Current density verification and thus the detection of electromigration issues are already an integral part of the sign-off verification of circuit layouts. Current density violations detected during sign-off are corrected by layout modifications—by the widening of wires, for example Lienig (2013).

Although there were many works aiming at mitigating the EM effects. The traditional EM analysis has focused on higher metal layers and in these types of interconnections: power supply network (Sect. 2.2.2), clock network (Sect. 2.2.3), and vias (Sect. 2.2.4). Thus, with the wire shrinking and increasing currents, the current densities in lower metal layers are also now in the range where EM effects are visible. So, the current flowing through the local interconnect wires within standard cells can be large enough to create significant EM effects over the lifetime of the chip. While the cell-internal signal EM problem is described in a patent (Domae and Ueda 2001) and industry publications (Jain and Jain 2012) as Sect. 2.2.6 presents, its efficient analysis is an open problem.

This work analyzes and gives a solution to improve the lifetime of the standard-cells considering the cell-internal signal EM problem.

Chapter 3
Modeling Cell-Internal EM

Gracieli Posser, Vivek Mishra, Palkesh Jain, Ricardo Reis, and Sachin
S. Sapatnekar

3.1 Modeling Time-to-Failure Under EM

For EM lifetime estimation, we use the well-known Black's equation (Black 1969)
developed by the physicist J.R. Black at the end of the 1960s:

$$\text{TTF} = A\, J^{-n} \exp\left(\frac{Q}{k_B T_m}\right) \tag{3.1}$$

where "TTF" is the time-to-failure (lifetime), A is an empirical constant that depends
on the material properties of the interconnect, J is the current density, the exponent n
is typically between 1 and 2, Q is the activation energy, k_B is Boltzmann's constant,
and T_m is the metal temperature. Black's equation shows that current density J and
the temperature T are deciding factors in the physical design process that affect
electromigration. Current density is giving by

$$J = \frac{I_{\text{avg}}}{T_w W}, \tag{3.2}$$

where W and T_w are the wire width and thickness and I_{avg} is the average current.

As 2011 ITRS presents (ITRS 2011), the lifetime of an interconnect is the time
to reach the minimum void size which is able to electrically open the interconnect.
As Kahng et al. (2013a) present, EM lifetime is affected by design parameters such
as wire width, fanout, driver size, and operating voltage—since all of these affect
current density. Switching activity (α), frequency, and temperature are the runtime

© Springer International Publishing AG 2017

G. Posser et al., *Electromigration Inside Logic Cells*,
DOI 10.1007/978-3-319-48899-8_3

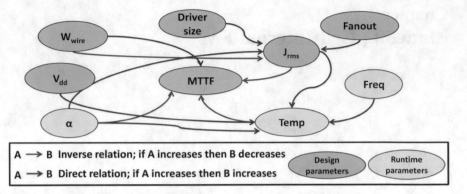

Fig. 3.1 Design and runtime factors affecting EM MTTF (Kahng et al. 2013a)

parameters that affect EM. Figure 3.1 (Kahng et al. 2013a) shows the relation to EM MTTF of these factors, with positive [negative] correlations to MTTF shown as red [blue] directed edges. For example, to meet EM MTTF margin at a given wire width upper bound we can reduce J_{rms} and driver size, producing a slower circuit. To meet EM MTTF margin at a given performance requirement, we can increase the wire width (W_{wire}) that increases the capacitance and dynamic power.

For wires with unidirectional current flow (e.g., many power grid wires), EM causes a steady unidirectional migration of metal items, and I_{avg} is simply the time-average of the current. In contrast, currents in signal wires may flow in both directions as they charge and discharge the output load.

For signal nets with bidirectional current flow, the time-average of the current waveform is often close to zero. However, even in cases where the current in both directions is identical, it is observed that EM effects are manifested. In this effect, often referred to as *AC EM*, the motion of atoms under one direction of current flow is partially, but not fully, negated by the "sweep-back" recovery effect that moves atoms in the opposite direction when the current is reversed. This partial recovery is captured by an *effective average current*, I_{avg} (Jain and Jain 2012; Lee 2012b) [also called Average Current Recovery (ACR) model (Ting et al. 1993)]:

$$I_{avg} = I_{avg}^+ - \mathscr{R} \cdot I_{avg}^-, \tag{3.3}$$

where $\mathscr{R}$ represents the *recovery factor* that captures sweep-back. Here, I_{avg}^+ is the largest of the average currents (forward-direction) and I_{avg}^- is the smallest current (reverse-direction). For signal wires in a cell, the rise and fall cycle currents are not always in opposing directions. We consider two cases:

Case I: When the rise and fall currents, I^r_{avg} and I^f_{avg}, are in opposite directions, as in edge e_3 in Fig. 1.5c, Eq. (3.3) yields:

$$I_{\text{avg}} = \frac{\max\left(\left|I^r_{\text{avg}}\right|, \left|I^f_{\text{avg}}\right|\right) - \mathscr{R} \cdot \min\left(\left|I^r_{\text{avg}}\right|, \left|I^f_{\text{avg}}\right|\right)}{2} \tag{3.4}$$

where the factor of 2 arises because half the transitions correspond to an output rise and half to an output fall.

In Ting et al. (1993), the ACR model heuristically accounts for the degree of damage recovery during opposite-polarity pulses through a single recovery parameter. The coefficient $\mathscr{R}$ accounts for the degree of damage recovery and may vary from zero to one. At $\mathscr{R} = 0$, any annealing or damage recovery due to opposite-polarity pulses is ignored, and the effective current density is due solely to the positive pulse contribution. At $\mathscr{R} = 1$, perfect annealing occurs during negative pulses.

Case II: When the rise and fall currents are in the same direction (e.g., in edge e_1 in Fig. 1.5c, where the charging rise current and the short-circuit current (not shown) during the fall transition both flow downwards), then

$$I_{\text{avg}} = \frac{\left|I^r_{\text{avg}}\right| + \left|I^f_{\text{avg}}\right|}{2} \tag{3.5}$$

In this work, we use a recovery factor $\mathscr{R}$ of 0.7, consistent with industry practice and published data (Lee 2012b). We use $A = 1.47 \times 10^7$ As/m^2 in SI units, which corresponds to an allowable current density of 10^{10} A/m^2 over a lifetime of 10 years at 378 K, with an activation energy, $Q = 0.85$ eV (Hu et al. 2007).

3.2 Joule Heating

The flow of current in an interconnect results in an increase in temperature within the wire due to a phenomenon known as Joule heating that accelerates temperature-dependent electromigration (Jonggook et al. 1999; Liu et al. 2015). From Eq. (3.1), a temperature rise hastens the EM-induced TTF. The temperature T_m in a wire is given by:

$$T_m = T_{\text{ref}} + \Delta T_{\text{Joule}} \tag{3.6}$$

where T_{ref} is the reference chip temperature for EM analysis and ΔT_{Joule} is the temperature rise due to Joule heating. In the steady-state, the wire temperature rises by Banerjee and Mehrotra (2001):

$$\Delta T_{\text{Joule}} = I^2_{\text{rms}} R R_\theta \tag{3.7}$$

We can observe that the wire temperature has an inherent squared dependency on the root mean square (RMS) wire current, I_{rms}, of the signal wire. Here, R is the wire resistance and R_θ is the thermal impedance of the wire to the substrate, giving by:

$$R_\theta = \frac{t_{ins}}{K_{ins}LW_{eff}}, \tag{3.8}$$

where t_{ins} is the dielectric thickness, K_{ins} is the thermal conductivity normal to the plane of the dielectric, L is the wire length, and W_{eff} is given by:

$$W_{eff} = W + 0.88t_{ins}, \tag{3.9}$$

for a wire width W. We obtain R by parasitic extraction using a commercial tool and use $t_{ins} = 120\,nm$ (FreePDK45 2011) and $K_{ins} = 0.07\,W/m\,K$ (Banerjee and Mehrotra 2001).

3.2.1   Local Hot Spots from Joule Heating (Li et al. 2014)

Li et al. (2014) present the importance of the Joule heating in the interactions among the neighboring metal lines. The Joule heating can rise the temperature in the local hot spots. Local hot spots can have significant impact on the number of critical EM elements.

One example of the neighboring metal line interactions is presented in Li et al. (2014) and it is illustrated in Fig. 3.2. Figure 3.2a shows a hypothetical layout block with signal lines only carrying AC current and power lines carrying DC current. Figure 3.2b illustrates the Joule heating effect on the center signal line (line #1) with different number of signal lines carrying AC currents to its I_{rms} limit, which is defined to cause a 5 °C temperature rise if the line is isolated. When all the signal lines carrying currents to their I_{rms} limits, the temperature rise from Joule heating (itself and its neighbors) of the central signal line can be greater than 25 °C, way above that if it is isolated.

Li et al. (2014) show that the similarly effect occurs with the interaction of the signal lines with the central VDD line, causing significant temperature rise in the VDD line as well. Though the current passing through the VDD line itself causes minimal joule heating, the neighboring signal lines acting together can heat the VDD line to higher than 10 °C, which could shorten its EM lifetime by almost 50 %. This 10 °C Joule heating on the center VDD line will increase its equivalent critical EM count from 1 (assuming it was designed to its J_{max} limit at the nominal use temperature) to 180,000. Evaluating the I_{avg} and I_{rms} interactions and local temperature from joule heating (i.e., identifying hot spots) is an important part of the chip level EM budgeting. This is especially true for the local areas tightly packed with actively switching signal and power lines (Li et al. 2014).

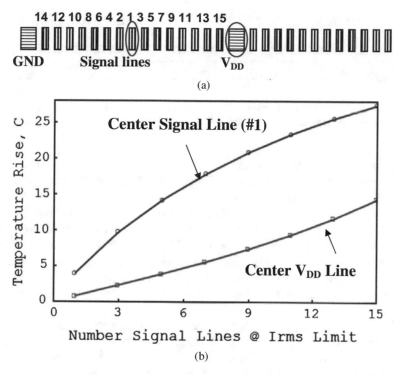

Fig. 3.2 An example of local Joule heating interactions caused by metal line temperature rise. The numbers used in the *x*-axis in (**b**) start from the closest neighbors of the subject line (#1 signal line or the center (*circled*)) VDD line (Li et al. 2014)

3.3 Current Divergence

In this work we are also considering the current divergence effect to calculate the effective average current through the edges of the signal wires. Current divergence addresses the EM interactions considering the vias and checks the EM reliability at the lead connection sites taking into account the lead EM interaction. The current divergence was presented in Park et al. (2010) and it is detailed below.

3.3.1 New Electromigration Validation: Via Node Vector Method (Park et al. 2010)

An EM check method named Via Node Vector is proposed in Park et al. (2010). This method addresses the EM interactions and checks the EM reliability at the lead connection sites (called via node), different than the conventional EM check

IN atomic flux (EM) = OUT atomic flux (EM)

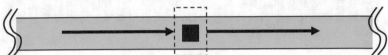

Fig. 3.3 As the "IN" and "OUT" atomic flux are the same, the atomic flux divergence along the interconnect is zero, resulting in no EM damage (Park et al. 2010)

that ignores the lead EM interaction and only checks the local current densities. It converts the electrical current density of each lead into an effective current density for the EM interaction consideration.

Electromigration (EM) is atomic flux driven by electrical current and its divergence at a site leads to interconnect failure, either in the form of increasing lead resistance or shorting out the circuits. The EM divergence is a net atomic flux at a site between IN and OUT electromigrated atoms. Even if the amount of atomic flux is very large along an interconnect, there can be no atomic flux divergence (EM damage) if the atomic flux is uniform along the interconnect, as Fig. 3.3 (Park et al. 2010) shows.

The local current density is related to the amount of the atomic flux but does not correctly represent the divergence of the atomic flux. So, Park et al. (2010) suggest that the best practice to check EM reliability is to limit the atomic flux divergence at a "via node" where the flux meets, such as vias, contacts, lead merge, or division points as the atomic flux is completely blocked and cannot migrate to the upper or lower layer.

A good example to explain the divergence problem given by Park et al. (2010) is presented in Fig. 3.4, where it shows the EM lifetimes of two connected leads, X and Y. The anode end of lead Y is connected to the cathode end of lead X which is stressed at $2j$, while lead Y is at j, where j is the current density. The width and length of the two leads are the same, thus $2\times$ more current flows through X than Y. The conventional method expects faster failure for lead X with $2j$, but experiments show that lead Y (with only j) fails $10\times$ earlier. The conventional check assumes the two leads are isolated in terms of EM reliability, but they are physically connected and their atomic flux indeed interacts. Thus, the actual divergence at the via node A and via node B is not purely determined by the local current density of the each lead X and Y, respectively. Lead X has a current flow of $2j$ but is supplied a current of j from lead Y. So, the electrical divergence at via node A is just j. Lead Y has a current flow of j, but the current keeps flowing into X with a larger amount of $2j$. The atomic flux from lead Y is not accumulated at via node A, but will be transferred into lead X at a faster rate due to the higher current density of $2j$. It will eventually increase the atom depletion rate from via node B and make the EM lifetime of lead Y short. A conservative electrical way to consider this effect is to add the two current

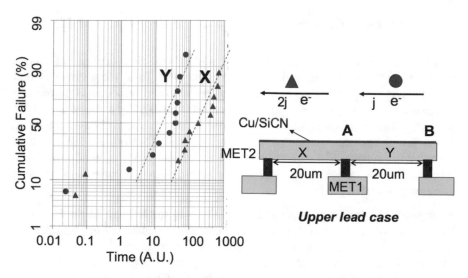

Fig. 3.4 Lead Y flows at half the current density of lead X, but fails faster than lead X; contrary to the conventional EM expectation. The new via node vector method explains it well with the divergence effect (Park et al. 2010)

densities, which makes the current divergence $3j$ from via node B, meaning that the EM lifetime of lead Y is similar to that of an isolated single lead having $3j$ current density. Thus lead Y fails much earlier than lead X. (The $3\times$ larger current density now makes $10\times$ shorter lifetime.)

The conventional EM check ignores the lead EM interaction in circuits and only checks the local current densities. The method presented in Park et al. (2010) addresses the EM interactions and checks the EM reliability at the lead connection sites (called via node). It converts the electrical current density of each lead into an effective current density for the EM interactions by applying three different non-electrical impacts on the atomic flux of a connected lead—length, width, and interaction—to electrical current density.

In this proposed work, the width is considered when we address the current density by $(I_{avg}/(WT_w))$, where I_{avg} is the average current, W is the width of the wire (interconnection), and T_w is the thickness of the wire. The length is influenced by the "back stress"[1] occurs in Al interconnects due to its protective oxide (Tu 2011), as in Cu interconnections the back stress force is not significant enough, we are not considering the length. And the interaction factor doesn't matter for our case. The way we are addressing the current divergence is better explained below.

---

[1] The back stress effect is proportional to JL of a lead, where J is the current density and L is a lead length. The lead will have a long EM lifetime when JL is reduced (Cheng et al. 2008).

3.3.2 Applying Current Divergence in the Proposed EM Model

A via in a copper interconnect allows the flow of electrical current but acts as a barrier for the migration of metal atoms under EM. Thus, the average current used for EM computation depends on the magnitude and direction of currents in neighboring wires where the metal migration flux is blocked by a via, as Sect. 3.3.1 presented. The computation of the average EM current can be performed according to the flux-divergence criterion presented in Park et al. (2010) (Sect. 3.3.1), which says that the average EM current for a wire is the sum of the current through the wire and the divergence at the via. *This new average current replaces all average currents in Sect. 3.1.*

Example. Consider the example of Fig. 3.5 showing the left half of the H-shaped INV_X4 (we can use any other logic gate) output wire presented in Fig. 1.5. Note that all metal wires within the H-shaped structure are routed on the same metal layer as in the Nangate 45 nm library (Nangate 2011), regardless of direction. Here, the output pin is placed at node 2 and consequently a via (in orange) is placed over this node. The arrows in Fig. 3.5 indicate the direction of electron flow of the current in this wire during the rise (in red) and fall (in green) transitions. Poly metal contacts (nodes 1, 3) are also blocking boundaries for metal atoms, and flux divergence must be used for wires at these nodes. Since voids in Cu interconnects are formed near the vias, we consider the two vias at either end of each edge. If an edge has multiple vias (e.g., e_1 has vias at nodes 1 and 2), the EM average current considering the divergence current, ($I_{\mathrm{avg},d}$), uses the largest divergence.

For edge e_1, node 1 does not see a void: the electron flow in this edge, during both the rise and fall transitions, is in the direction of node 1, and EM voids are only caused by electron flow away from the via. However, for the via at node 2, there is an effective outflow and the EM average current for edge e_1 with respect to via 2,

Fig. 3.5 Current divergence example considering a multifanout tree

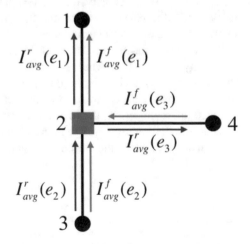

$I_{\text{avg},d}(e_1)$, is computed using Eq. (3.5) as the rise and fall electron flows are in the same direction:

$$I_{\text{avg},d}(e_1) = \frac{\left| I^r_{\text{avg},d}(e_1) \right| + \left| I^f_{\text{avg},d}(e_1) \right|}{2}$$

where

$$I^r_{\text{avg},d}(e_1) = I^r_{\text{avg}}(e_1) - I^r_{\text{avg}}(e_2) + I^r_{\text{avg}}(e_3)$$

$$I^f_{\text{avg},d}(e_1) = I^f_{\text{avg}}(e_1) - I^f_{\text{avg}}(e_2) - I^f_{\text{avg}}(e_3)$$

The expression for $I^r_{\text{avg},d}$ above has contributions from:

- Current in e_1, drawing metal flux away from the via, and adds to void formation.
- Current in e_2, which inserts flux into the via: although this current flows to the output load through the via at node 2, due to the blocking boundary at the via, the metal flux does not pass through, but instead, accumulates atoms, thus negating void formation.
- Current in e_3, which draws flux away from node 2.

The expression for $I^f_{\text{avg},d}$ is similarly derived.

For edge e_3, we see Case II of Sect. 3.1 because the rise and fall electron flows are in the opposite direction. From Eq. (3.4),

$$I_{\text{avg},d}(e_3) = \frac{\left| I^r_{\text{avg},d}(e_3) \right| - \mathcal{R} \cdot \left| I^f_{\text{avg},d}(e_3) \right|}{2}$$

where

$$I^r_{\text{avg},d}(e_3) = I^r_{\text{avg}}(e_3) - I^r_{\text{avg}}(e_2) + I^r_{\text{avg}}(e_1)$$

$$I^f_{\text{avg},d}(e_3) = I^f_{\text{avg}}(e_3) + I^f_{\text{avg}}(e_2) - I^f_{\text{avg}}(e_1)$$

The Algorithm 3.1 presents the current divergence calculation, where all vias (vias and contacts) v are visited. For each via v, the neighbor edges e are stored in $nbr_edges(v)$. For each edge e in $nbr_edges(v)$, the $I^r_{\text{avg},d}$ and $I^f_{\text{avg},d}$ are calculated as example presented. The $I_{\text{avg},d}$ of the edge is calculated using Eq. (3.4) when the rise and fall electron flows are in the opposite direction and Eq. (3.5) when the electron flows are in the same direction. Finally, the $I_{\text{avg},d}$ value of the edge considering that via is stored. After all vias and all edges be visited, each edge e is visited again. The $I_{\text{avg},d}$ for each edge e will be the largest divergence among all vias.

Algorithm 3.1 Current Divergence Calculator

    V = List of all Vias
    $nbr\_edges(v)$ = List of neighboring edges in V
  1: **for** each v in V **do**
  2:     Get neighbors of v store it in $nbr\_edges(v)$
  3:     **for** each edge e in $nbr\_edges(v)$ **do**
  4:         Calculate $I^r_{avg,d}, I^f_{avg,d}$ using (Park et al. 2010)
  5:         Calculate $I_{avg,d}$ using Eq. (3.4) or Eq. (3.5)
  6:         Store this value as $I_{avg,d}(e, v)$
  7:     **end for**
  8: **end for**
  9: **for** each edge e **do**
10:     $I_{avg,d} = \max_{i \in V}(I_{avg,d}(e, i))$
11: **end for**

3.3.3 The Impact of Blech Length on Cell-Internal Interconnects

As pointed out by Blech (1976), the migration of metal atoms results in a concentration gradient and a back stress that opposes the electron wind force that causes electromigration. This is typically translated into a criterion whereby the product of the current density and wire length must exceed a critical value; if it does not, the wire is deemed immortal.

Although intracell wires are short, the Blech criterion cannot be applied directly to signal interconnects in standard cells, as indicated in Jain and Jain (2012) and Lee (2012b). This may also be explained by observing that the bidirectional nature of AC EM does not allow a substantial build-up in the concentration gradient, and therefore the back stress that opposes the electron wind is limited. For Vdd and Vss wires, although the wires shown in an individual cell may be short, they are typically concatenated along an entire row of standard cells, implying that the actual length of the wire is much larger than the short segment seen in the layout schematic of a single cell, to the point where the length of the wire does not make it immortal under the Blech criterion.

3.4 Conclusions

In this section we presented how we are modeling the cell-internal EM. To calculate the lifetime we are using the Black's equation (Eq. (3.1)), where the current density J is calculated using the AVG current (I_{avg}). When the rise and fall currents are in opposite directions, I_{avg} is calculated as in Eq. (3.4), using the recovery factor. When the currents are in the same direction, I_{avg} is calculated using Eq. (3.5).

The Joule heating effects are incorporated in the lifetime calculation using Eq. (3.7). Moreover, the current divergence is also considered in our work on the computation of the average EM current and consequently computing the lifetime, as Sect. 3.3 presented.

Chapter 4
Current Calculation

Gracieli Posser, Palkesh Jain, Vivek Mishra, Ricardo Reis,
Sachin S. Sapatnekar

This chapter presents how the average and RMS current values are calculated to model the EM effects in this work. *The current modeling has also the contribution of two other authors: Vivek Mishra (from University of Minnesota) and Palkesh Jain (from Qualcomm India).*

For a standard cell with m output pin positions, characterization for delay and power can be performed at any one of the pin positions. Since the cell-internal wire parasitics in a standard cell are negligible and are dominated by transistor parasitics, this characterized value is accurate at all other pin locations and practically does not affect the cell timing. This is also true for the transients on the Vdd and Vss pin networks, which are essentially independent of the pin positions.

The evaluation of EM TTF requires a characterization of (a) the average and RMS currents through a Vdd/Vss line and (b) the average currents, I^r_{avg} and I^f_{avg} and the RMS current I_{rms}. All of these parameters are dependent on the pin position, as demonstrated in Fig. 1.5, and an obvious approach would be to enumerate the characterization over all possible combinations. For a library with N_{lib} cells, each with an average of m output pin positions, d Vdd pin positions and s Vss pin positions, this implies $m \times d \times s$ characterizations. However, given that EM evaluations in Vdd, Vss, and signal nets are independent, this can be brought down to $m + d + s$ characterizations. With this reduction, the CPU time required for standard cell characterization is given by:

$$T_{char} = (m + d + s) \cdot N_{corners} \cdot N_{lib} \cdot T^{avg}_{char,cell} \tag{4.1}$$

where $N_{corners}$ represents the number of corners at which the cell is characterized, and $T^{avg}_{char,cell}$ is the average characterization time (typically SPICE simulations for the output rising/falling cases) for each cell. A typical library may have $N_{lib} = 200$ cells. In our experiments, the average characterization time to build the 7×7 .lib table for a cell in the 45nm NANGATE library is found to be $T_{char,cell}^{avg} = 17.5$s. For the NANGATE library, the average number of pin positions $m = 12$, $d = 10$, $s = 10$, and the number of corners, $N_{corners} = 15$. This yields $T_{char} = 19$ days to characterize the library, which is $(m + d + s)$ times (=32× for this example) the cost of characterizing each cell at one pin position for output, Vdd and Vss pins. At more advanced process nodes, the number of corners goes up significantly, and therefore T_{char} can be much higher if the corners characterization are not parallelized.

In this work, we show that a simpler approach is possible, speeding up the characterization time by a factor of almost $(m + d + s)$, 32 for our example above. This implies that the above 19-day characterization can be conducted more practically, in about half a day. Our procedure extracts the average and RMS current information from the same simulations used for delay and power characterization, at a *reference pin position*, and then uses inexpensive graph traversals to evaluate EM for other pin positions. In other words, the additional overhead over conventional cell characterization is negligible.

To illustrate the EM characterization procedure for the output signal wire, consider INV_X4 in Fig. 4.1 with the output pin at node 4. We will temporarily ignore short-circuit and leakage currents to simplify the example. Here, all PMOS [NMOS] devices are identical and inject equal charge/discharge currents. Figure 4.2 presents an example using the average current values for the rise transition from a SPICE simulation. Figure 4.2(a) presents the values when the output pin is at node 4 (reference pin position). When the pin is moved to node 2 [node 6], the distribution of currents in the branches remains similar, except edge e_3 [e_4], which now carries an almost equal current in the opposite direction. Figure 4.2(b) shows the average current values when the output pin is moved to node 2, where just the current through the edge e_3 changes, the values and arrows in grey show the unchanged values and current directions when the pin is moved. The same happens when the pin is at node 6, as Fig. 4.2(c) shows. Therefore, the Joule heating and EM lifetime for each edge are unchanged, and only the current divergence calculations change because the current direction changed.

When the pin is moved from node 4 to node 3, the PMOS current injected at node 5 is redirected to also flow through e_2 and e_3. The only changed current magnitudes correspond to segments e_2 and e_3; those for the other wire segments remain almost the same since intracell wire parasitics are small, as Fig. 4.3 shows.

Both cases above show incremental changes in current flow patterns when the pin is moved. Similar observations may be made when the Vdd and Vss pins are moved: in each case, the difference from moving a pin arises because of a redirection of a set of currents. These facts indicate that it may be possible to reduce the characterization effort by performing a single SPICE simulation for one pin position, called the *reference case*, and inferring the current densities for every other pin position from this data by determining the current redirection. We develop a graph-based method for determining this redirection, and an algebra for computing I_{avg} and I_{rms} for each pin position based on the values from the reference case.

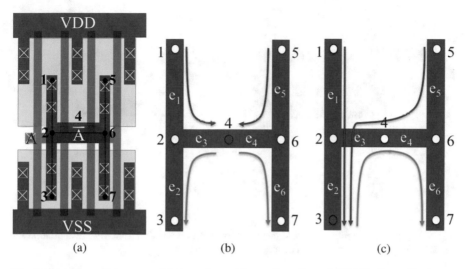

Fig. 4.1 (**a**) The cell layout and output pin position options for an INV_X4. Charge/discharge currents when the output pin is at (**b**) node 4 and (**c**) node 3. The red [green] lines represent rise [fall] currents.

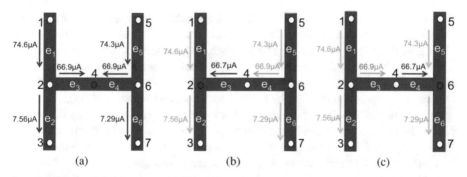

Fig. 4.2 (**a**) AVG current values in Ampere (A) for the rise transition from SPICE simulation for the INV_X4 when the output pin is at node 4, (**b**) node 2 and (**c**) node 6.

The reference case is characterized for a fixed reference frequency, f_{ref}, chosen to be 1GHz in our experiments. If a given design operates at a frequency f and an activity factor α, as long as the circuit operates correctly at that frequency (i.e., all transitions can be completed), it is easy to infer the average and RMS currents in each branch. The average and RMS currents are multiplicatively scaled by factors of $\alpha f/f_{ref}$ and $\sqrt{\alpha f/f_{ref}}$, respectively.

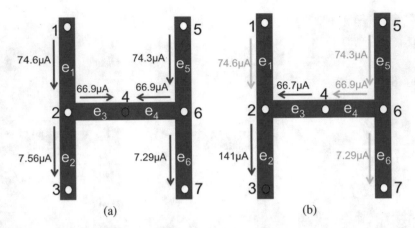

Fig. 4.3 (**a**) AVG current values in Ampere (A) for the rise transition from SPICE simulation for the INV_X4 when the output pin is at node 4 and (**b**) node 3.

4.1   Current Flows Using Graph Traversals

We present a graph-based algorithm that computes the currents through each edge when the pin position is moved from the reference case to another location. Our algorithm captures the effect of charge/discharge currents, short-circuit currents, and leakage currents (neglected in the example above), and its pseudocode is shown in Algorithm 4.1. For the output net (but not for Vdd/Vss nets), the short-circuit and leakage currents are unaffected by the pin location, and for all nets, Figs. 4.1, 4.2 and 4.3 show that the flow of the charge/discharge currents is affected by the output pin position. Coupling capacitance currents are the same for almost all nets since moving the pin does not significantly change the transient waveforms in these nets. Appendix B explain why the coupling capacitances currents are also considered.

Our algorithm uses graph traversals to trace the change in the current path when the pin position is moved from the reference pin position, *ref*, to any candidate pin position on the output net, as enumerated in a candidate set C. Lines 1–6 perform a SPICE simulation at reference pin location *ref* to compute each average and triangle representations for edge currents during rise and fall on the output net, and over the cycle for the Vdd and Vss components. The charge/discharge, short-circuit/leakage and coupling capacitance currents for each edge are determined from the simulation.

The output metallization has several points that are connected to the NMOS and PMOS transistors: we refer to these as *current injection points*. For example, in Fig. 1.5(b), the NMOS and PMOS current injection points are at nodes $\{1,5\}$ and $\{3,7\}$, respectively. Next, in the **for** loop that commences at line 7, we determine the current contribution for each candidate pin position in C. The graph-based approach determines the unique path P_i from the reference pin position *ref* to pin candidate i (line 8). Note that the Vss pin draws current out from the cell while the Vdd pin injects current, and therefore the direction of the path P_i is reversed for the

---

Algorithm 4.1 Efficient cell EM current characterization.

Input: Undirected graph $G(V, E)$ with separate connected components for the cell output, Vdd,
   and Vss nets; Reference pins *ref* for output, Vdd, and Vss for each connected component $\in V$;
   Set of candidate pin positions $C \subseteq V$ for output, Vdd, and Vss components.
Output: I_{avg} for all Vdd and Vss edges, $I_{avg}^{+}(e)$, $I_{avg}^{-}(e)$ for all output edges, $I_{rms}(e) \; \forall \; e \in E \; \forall$
   pin positions in C.
 1: SPICE-simulate the cell with the output, Vdd, and Vss at *ref*, find triangle representations,
    average of edge currents during rise, fall.
 2: **for each** connected components $\in$ Vdd, Vss, output **do**
 3:     **for each** current injection point j **do**
 4:         $P_j^{\{r/f\}} = \{$charge/discharge$\}$ path from j to *ref*.
 5:         Find charge/discharge, short-circuit/leakage, and coupling capacitance currents
    injected at j.
 6:     **end for**
 7:     **for each** pin position $i \in C$ **do**
 8:         Compute unique path P_i from *ref* to pin position i. The direction of P_i is from *ref* to i
    for output and Vss, and from i to *ref* for Vdd.
 9:         **for each** current injection point j **do**
10:             New $\{$charge,discharge$\}$ path from j to i, $P_j'^{\{r/f\}}$ = algebraic sum of paths P_i and
    $P_j^{\{r/f\}}$.
11:             Update the current for each edge in P_j', For the output net, update only the
    $\{$charge,discharge$\}$ current, keeping short-circuit/leakage, and coupling capacitance currents
    unchanged; for Vdd/Vss nets, update all currents, except coupling capacitance currents which
    are unchanged.
12:         **end for**
13:         Compute $I_{avg}(e)$ for Vdd/Vss or $\{I_{avg}^{+}(e), I_{avg}^{-}(e)\}$ for output, as well as $I_{rms}(e) \; \forall \; e \in E$
    for pin position i.
14:     **end for**
15: **end for**
16: **return**

---

two cases. For the output pin, we use the same direction as the Vss pin, but the
precise direction does not matter due to the max/min operators used in Eq. (3.4).

For each current injection point, the charge/discharge path for pin candidate i
(lines 9–12) is the algebraic sum of P_i and the charge/discharge path P_j for the
reference pin position. The currents are updated in line 13.

Example (output pin): The key idea is illustrated in Fig. 4.4 for the rise transition
when the pin is moved from reference node 4 to node 3: the unique path P_3 between
these nodes is shown at left. The two figures on the right show the algebraic addition
of path P_3 with paths P_1^r and P_5^r, respectively, corresponding to the two rise current
injection points. After cancellations, the resulting path successfully shows the new
path for charging currents: $\{e_1, e_2\}$ for the PMOS current from node 1, and $\{e_5,
e_4, e_3, e_2\}$ for the PMOS current from node 5. The charge/discharge currents are
updated in lines 9–11, while the short-circuit and leakage contributions are the same
as the reference case.

Example (Vdd pin): Figure 4.5 shows an example of how our graph-based algorithm
is applied for the Vdd pin. The example considers the case when the pin is moved

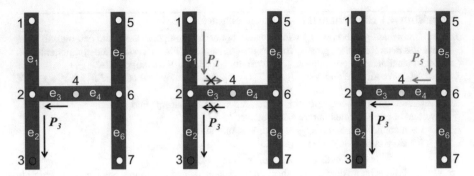

Fig. 4.4 Recomputation of the rise currents when the pin is moved from node 4 (reference node) to node 3 (new position).

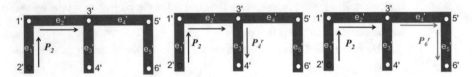

Fig. 4.5 Our graph based algorithm applied to the Vdd pin when the pin is moved from node $3'$ to node $2'$.

from reference node $3'$ to node $2'$: the unique path P_2 between these nodes is shown at left. Note that according to line 6, the direction of this path is the opposite of that used for the output and Vss components. The algebraic addition of path P_2 with paths P_4^r and P_6^r is shown on the two figures on the right, respectively, corresponding to the two rise current charging points. The resulting paths for charging currents are: $\{e_3', e_2', e_1'\}$ for the PMOS current from node $4'$, and $\{e_5', e_4', e_2', e_1'\}$ for the PMOS current from node $6'$.

4.2   Algebra for Average/RMS Current Updates

The current waveforms in the wire segments, for the rise and fall transitions, are used to calculate the RMS and effective average current through the wire: the former is used to measure self-heating, and the latter is used in the EM TTF formula. We now develop an algebra for efficient RMS and effective average current updates for various pin positions, given information for the reference case.

4.2.1 Algebra for Computing Average Current

For edge e, I_{avg} during a rise or fall half-cycle is given by:

$$I_{avg}(e) = \frac{1}{T/2} \int_0^{T/2} I(e)(t)dt = \frac{1}{T/2} \sum_{i \in S} \int_0^{T/2} I(p_i(e))(t)dt \qquad (4.2)$$

where the summation is over the set S of all current insertion points whose currents contribute to the current in edge e.

When the pin is moved, the set S is modified, and some entries are added and removed to the set. For example, in Fig. 1.5, when the pin is moved from node 4 to node 3, the current in edge e_2 has new contributions from current insertion points 5 (rise) and 7 (fall) and a removal of the contribution from point 3; the current in e_3 must subtract the contribution of current insertion point 1 (rise) and 3 (fall), and add contributions from insertion points 5 (rise) and 6 (fall). To perform these operations, we can simply add or subtract the average currents associated with the corresponding current insertion point. For a current $I(p_i)$ from a pin insertion point p_i that is added or subtracted, we can write

$$(I(e) \pm I(p_i))_{avg} = \frac{1}{T/2} \int_0^{T/2} (I(e)(t) \pm I(p_i))dt$$

$$= I_{avg}(e) \pm I_{avg}(p_i)$$

Therefore, I_{avg} updates for a new pin position simply involve add/subtract operations on average reference case currents.

Example: For the Vdd net example shown in Fig. 4.5, we illustrate how the average current values are updated. Figure 4.6 a, b show the formal representation of how the currents change when the pin is moved from node 3′ to node 2′, while Fig. 4.6 c, d show the SPICE simulation results when the pin is at node 3′ and at node 2′, respectively. Figure 4.6(a) shows the rise currents $I_{avg}(p_i)$ charging the pin insertion points p_i when the Vdd pin is placed at node 3′. In this example, for the INV_X4, there are three insertion points, 2′, 4′ and 6′. When the pin is moved from node 3′ to node 2′, the currents through the edges e'_3, e'_4 and e'_5 remain the same and are shown in a grey color in Fig. 4.6 b, d. The currents that must be updated are those through the edges e'_1 and e'_2, $I_{avg}(e'_1)$ and $I_{avg}(e'_2)$, respectively. Calculating by our algebra and using the notation and values in Fig. 4.6, $I_{avg}(e'_1)$ and $I_{avg}(e'_2)$ are each given by:

$$I_{avg}(e'_1) = I_{avg}(e'_2) = I_{avg}(p_4) + I_{avg}(p_6) = 27.6\mu A + 13.8\mu A = 41.4\mu A$$

Comparing the calculated by our algebra with the value from SPICE simulation for the new pin position, this value is seen to be very close to the actual value of $42\mu A$.

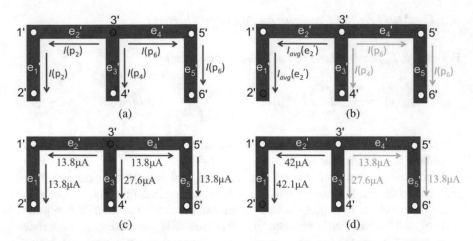

Fig. 4.6 Vdd pin position options for INV_X4 and the current values when the Vdd pin is at node 3' ((a) and (c)) and node 2' ((b) and (d)).

Example for the output pin: Using the AVG current values (in Ampere) in Fig. 4.3, when the pin is moved from node 4 to node 3, the current in edge e_2 has the contribution from the current insertion point 1 (74.6μA) and a new contribution from current insertion point 5 (74.3μA) and a removal of the contribution from insertion point 3 (7.56μA). Then, the current through e_2 when the pin is at node 3 is giving by 74.6μA + 74.3μA - 7.56μA = 141μA, as Fig. 4.3(b) shows. The current in e_3 must subtract the contribution of current insertion point 1 (74.6μA), and add contributions from insertion point 5 (74.3μA), so the current through e_3 is giving by 66.9μA - 74.6μA + 74.3μA = 66.6μA, practically the same as computed by SPICE simulation (as shown in Fig. 4.3(b)) that was 66.7μA.

4.2.2   Algebra for Computing the RMS Current

The waveform for the current drawn by each device may be approximated by a triangle with height I_a, and with a nonzero current for a period of T' seconds, where $T' < T$, the clock period (this current model is widely used). It is well-known (Nastase 2013) that the RMS value of such a waveform is

$$I_{rms,\Delta} = I_a \sqrt{\frac{T'}{3T}} \tag{4.3}$$

Due to the tree structure of the output wire, the current in each edge is a sum or difference of a set of such triangular signals, and this set can be determined based on a tree traversal. The sum (or difference) of a set of triangular waveforms, potentially each with different heights, start times, and end times, can be represented as a

Fig. 4.7 The sum of the two upper triangular waveforms (waveform 1 and 2) can be represented as a set of piecewise triangular or trapezoidal segments (resulting waveform).

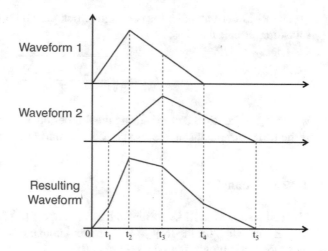

piecewise linear waveform, and thus each edge current has this form. To find the RMS value of such a piecewise linear waveform, we can decompose it into a set of nonintersecting (except at the edges) triangles and trapezoids, as shown in Fig. 4.7.

The RMS for this waveform can be shown to be:

$$I_{rms}^2 = \sum_{\text{all triangles } i} I_{rms,\Delta_i}^2 + \sum_{\text{all trapezoids } i} I_{rms,trap_i}^2 \tag{4.4}$$

To use the above equation, we use Eq. (4.3) for the RMS of a triangular waveform, and the following formula for the RMS of a trapezoid bounded by the time axis, with value I_b at time b and I_c at time c, where $c > b$:

$$I_{rms,trap} = \sqrt{\frac{\left(I_b^2 + I_b I_c + I_c^2\right)(c - b)}{3T}} \tag{4.5}$$

For INV_X4, since the transistors of each type are all identical and are driven by the same input signal, each PMOS [NMOS] device injects an identical charging [discharging] current waveform; however in general, the currents may be different. Since the intracell resistive parasitics of the output metallization are small, some combination of these nearly unchanged currents is summed up along each edge during each half-cycle. The set of triangular PMOS waveforms that contribute to the current in each edge in Fig. 4.1 is simply the set of PMOS devices i whose charge or discharge path (Algorithm 4.1) traverses edge i. When the output is moved from node 4 to node 3, the current through an edge loses some set membership and gains others. The updated set of triangles add up, in general, to a waveform with triangles and trapezoids, whose RMS value is given by Eq. (4.4). For the Vdd and Vss rails, the currents are updated in the same way. Vdd rail injects current to charge the PMOS devices and the Vss rail discharge the current from the NMOS transistors.

The RMS current value for the entire clock period considering the rise and fall transitions is calculated as:

$$I_{rms} = \sqrt{\frac{(I_{rms}^f)^2 + (I_{rms}^r)^2}{2}} \tag{4.6}$$

where I_{rms}^f is the RMS current during the fall period, i.e., during half period and I_{rms}^r is the RMS current during the rise period, i.e., during the other half period.

4.2.2.1   Example

Figure 4.8 shows the RMS current values in Ampere (A) for the rise transition from SPICE simulation for the INV_X4 when the output pin is at node 4 (Fig. 4.8(a)) and when the pin is moved to node 3 (Fig. 4.8(b)).

To calculate the RMS current, we will consider the same idea used to calculate the average current, considering the Algorithm 4.1 where a pin position is used as reference case to calculate the current for the other pin positions. In this way, the RMS current in edge e_2 has the contribution from the current insertion point 1 and a new contribution from current insertion point 5 (point 1 and point 5 have the same current value) and a removal of the contribution from insertion point 3. So, the current through e_2 when the pin is at node 3 is given by $Wave_{point1} + Wave_{point5} - Wave_{point3}$, i.e., the addition of the waveforms of the insertion point 1 and 5 ($Wave_{point1} + Wave_{point5}$) and the subtraction of the waveform of the insertion point 3 ($Wave_{point3}$).

Figure 4.9 presents the current waveforms from a SPICE simulation for the rise transition of the insertion point 1 (top wave), that is equal to the waveform of the

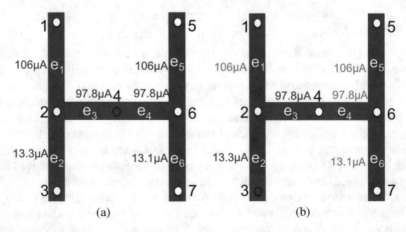

(a)   (b)

Fig. 4.8 (a) RMS current values in Ampere for the rise transition from SPICE simulation for the INV_X4 when the output pin is at node 4 and (b) node 3.

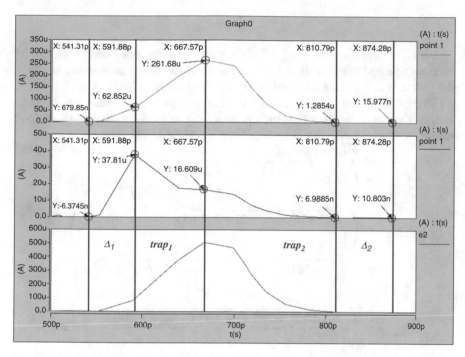

Fig. 4.9 Current waveforms from a SPICE simulation for the rise transition of the insertion point 1 (that is equal of the wave for the insertion point 5) and insertion point 3 considering that the pin position is at the reference case (pin at node 4). The bottom wave is the resulting wave representing the current through e_2 when the pin is moved to node 3 (from (Synopsys 2013a) and adapted by the author (2015)).

insertion point 5, and insertion point 3 (middle wave) for the reference case of the pin position (pin at node 4). When the pin is moved to node 3, the resulting wave for the current through e_2 is the bottom wave in the Fig. 4.9.

To compute the RMS current through e_2, we start computing the I_{rms} for each triangle and trapezoid of the waves, using the eq. 4.3 and 4.5. The I_{rms} through the edge e_2 considering the waves in Fig. 4.9 will be given by:

$$I_{rms_{e2}}^{rise\ 2} = I_{rms,\Delta_1}^2 + I_{rms,trap_1}^2 + I_{rms,trap_2}^2 + I_{rms,\Delta_2}^2$$

where

$$I_{rms,\Delta_1}^2 = I_{rms,\Delta_{1_{point1}}} + I_{rms,\Delta_{1_{point5}}} - I_{rms,\Delta_{1_{point3}}}$$

$$I_{rms,trap_1}^2 = I_{rms,trap_{1_{point1}}} + I_{rms,trap_{1_{point5}}} - I_{rms,trap_{1_{point3}}}$$

$$I_{rms,trap_2}^2 = I_{rms,trap_{2_{point1}}} + I_{rms,trap_{2_{point5}}} - I_{rms,\Delta_{2_{point3}}}$$

$$I^2_{rms,\Delta_2} = I_{rms,\Delta_{2_{point1}}} + I_{rms,\Delta_{2_{point5}}}$$

Using the values shown in Fig. 4.9 of the waves for point 1 and point 3, the I_{rms} for each triangle and trapezoid is calculated. As the simulation was executed at 1GHz, the period is 1ns and the half period for the rise transition is 0.5ns, so T is 0.5ns. As the current of the injection point 5 is equal to point 1, the values are the same and, so, the values for point 1 are multiplied by 2. The units are not in the equations, where the current values are represented in μA and the time values are in ns.

$$I^2_{rms,\Delta_1} = \left(62.852\sqrt{\frac{0.051}{1.5}}\right) * 2 - \left(37.81\sqrt{\frac{0.051}{1.5}}\right)$$

$$I^2_{rms,trap_1} = \left(\sqrt{\frac{(62.852^2 + 62.852 * 261.68 + 261.68^2)(0.076)}{1.5}}\right) * 2$$

$$- \left(\sqrt{\frac{(37.81^2 + 37.81 * 16.609 + 16.609^2)(0.076)}{1.5}}\right)$$

$$I^2_{rms,trap_2} = \left(\sqrt{\frac{(261.68^2 + 261.68 * 1.285 + 1.285^2)(0.143)}{1.5}}\right) * 2$$

$$- \left(16.609\sqrt{\frac{0.143}{1.5}}\right)$$

$$I^2_{rms,\Delta_2} = \left(1.285\sqrt{\frac{0.063}{1.5}}\right) * 2$$

resulting in

$$I^2_{rms,\Delta_1} = (11.59 * 2) - (6.97) = 16.21$$

$$I^2_{rms,trap_1} = (67.10 * 2) - (10.87) = 123.33$$

$$I^2_{rms,trap_2} = (81.00 * 2) - (5.13) = 156.87$$

$$I^2_{rms,\Delta_2} = (0.26 * 2) = 0.52$$

So, the final result will be:

$$I^{rise}_{rms_{e2}}{}^2 = (16.21^2) + (123.33^2) + (156.87^2) + (0.52^2)$$

$$I^{rise}_{rms_{e2}} = \sqrt{262.76 + 15210.29 + 24608.2 + 0.27}$$

$$I^{rise}_{rms_{e2}} = 200.20\mu A$$

The $I^{rise}_{rms_{e2}}$ result given by our formulation, 2e-4A, is very close to the result given by SPICE simulation, 2.03e-4A, shown in Fig. 4.8. So, we can see that the formulation works as expected.

The current in e_3 must subtract the contribution of current insertion point 1 and add contributions from insertion point 5, so the current through e_3 is given by $Wave_{e_3} - Wave_{point1} + Wave_{point5}$, as the current values for the $Wave_{point1}$ and $Wave_{point5}$ are the same, the RMS current value through e_3 does not change when the pin position is changed from node 4 to node 3. In Fig. 4.8 the values are a little bit different, 9.78e-5A and 9.75e-5, but they are very close, keeping the accuracy of our results.

4.3 Results

Table 4.1 shows the results of our characterization approach for the set of cells of the NANGATE cell library used in this work based on a single SPICE simulation, followed by graph traversals and the current update algebra. These results were calculated for the output pin positions. One reference case is chosen for each cell and the number of output candidate pin positions varies from 6 to 25, with an average of about 12 pin candidates per cell. The number of pin candidates is the number of simulations saved using our graph-based algorithm.

The number of Vdd candidate pin positions varies from 4 to 26 and for Vss varies from 5 to 26, with an average of about 10 pin candidates per cell, as Tables 6.3 and

Table 4.1 Comparison with SPICE for I_{avg} calculated using our algorithm

Cell	# Candidates	SPICE	Ours	Error (%)
NAND2_X2	8	4.72e-5	4.70e-5	0.32
NAND2_X4	10	4.27e-5	4.31e-5	0.99
NOR2_X2	6	2.74e-5	2.76e-5	0.72
NOR2_X4	8	2.22e-5	2.23e-5	0.28
AOI21_X2	8	3.81e-5	3.81e-5	0.09
AOI21_X4	11	3.00e-5	2.96e-5	1.23
INV_X4	7	9.84e-5	9.88e-5	0.46
INV_X8	13	1.02e-4	1.02e-4	0.64
INV_X16	25	1.29e-4	1.28e-4	0.63
BUF_X4	7	9.79e-5	9.85e-5	0.57
BUF_X8	13	1.12e-4	1.11e-4	0.36
BUF_X16	25	1.24e-4	1.25e-4	0.08
AVG	11.8	-	-	**0.53**

For each cell, the value corresponds to the edge current with the largest error

6.4 show. For this library, the number of SPICE simulations is therefore reduced by about $32\times (12+10+10)$, significant and worthwhile savings even for an one-time library characterization task. Table 4.1 shows the edge within each cell that shows the largest error for the effective average current: in each case, this error is seen to be small, 0.53% on average, while the computational savings for characterization are large. The largest error is 1.23% and the smallest error is 0.08%.

Chapter 5
Experimental Setup

We now present the experimental setup used in this work for analyzing and improving circuit lifetime under cell-internal EM. Since we do not have yet access to a library at a recent technology node, where EM effects are more significant (Jain and Jain 2012), our evaluation is based on scaling layouts from the NANGATE 45 nm cell library down to 22 nm. While this may not strictly obey all design rules at a 22 nm node, the transistor and wire sizes are comparable to 22 nm libraries, and so are the currents. The layout parasitic extraction was done using the 45 nm FreePDK (FreePDK45 2011) models and the Calibre xRC (Mentor 2013) tool.

Initially the cells are characterized for the average and RMS currents in each cell under a reference pin position. The cells are characterized considering $f_{ref} = 1$ GHz and for seven different values each for the input slew and output load. As higher is the frequency larger will be the EM effects. The characterization thus generates a 7×7 look-up table with the RMS and average current values for the slew and load values, and these values are determined based on SPICE characterization of the scaled 22 nm library based on publicly available 22 nm SPICE ASU PTM models for the High Performance applications (PTM HP) (Zhao and Cao 2007).

Hereafter, the analysis follows the flow presented in Fig. 5.1. First, we synthesize ITC'99 and ISCAS'89 benchmarks using Design Compiler (Synopsys 2013b) with delay specs set to the best achievable frequency. The cells from the NAN-GATE library (Nangate 2011) used in this work are: NAND2_X2, NAND2_X4, NOR2_X2, NOR2_X4, AOI21_X2, AOI21_X4, INV_X4, INV_X8, INV_X16, BUF_X4, BUF_X8, BUF_X16, DFF_X2, DFFR_X2, and DFFS_X2. We focus on EM in the combinational cells.

Each circuit is placed and routed using Cadence Encounter (Cadence 2013). After the routing, the SPEF file with the extracted wire RCs and the Verilog netlist are saved. The timing, power, area, and wirelength are reported. Synopys PrimeTime reads the SPEF, Verilog, and SDC files and reports the input slew, output load, and switching probability (α) for each instance of the circuit. The PrimeTime timing report provides the slew, load, and switching probability for all cell instances. These

© Springer International Publishing AG 2017
G. Posser et al., *Electromigration Inside Logic Cells*,
DOI 10.1007/978-3-319-48899-8_5

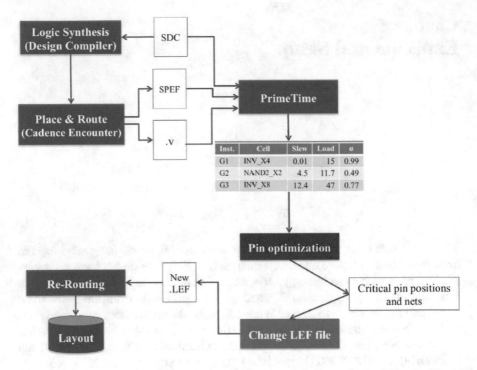

Fig. 5.1 The implementation flow used in this work considering the standard-cell based synthesis flow

informations are used as input in our method to optimize the TTF (lifetime) of the circuits by the output pin position optimization, where its flow is shown in Fig. 5.2.

To optimize the pin position considering all instances in the circuit, for each instance, based on the reported slew and load, we calculate I_{avg} and I_{rms} for each internal wire segment of the output pin (e.g., edges from e_1 to e_6 in Fig. 4.1), interpolating from a 7×7 look-up table characterized for the reference pin position, and infer currents for each candidate position using the approach of this work presented in Chap. 4. The TTF is found using Eq. (3.1) at 378 K, a typical EM specification. To calculate the TTF, beyond the I_{avg} and I_{rms} values, the wire information (width, length, and resistance) are also important and are provided considering the dimensions of the cell layout and the parasitic extraction. The TTF for each wire segment (edge) is calculated considering these information.

The **worst TTF** of the circuit is given by the instance in the circuit that has the smallest TTF. For this, for each instance is calculated the best and the worst TTF, i.e., the best and the worst pin position. The worst TTF for each pin position is the internal wire with the smallest TTF. Considering that the failure of a single wire caused by EM can result in the failure of the entire circuit (Lienig 2013). To compute the best TTF that the circuit can achieve under output pin selection, for each cell we determine the output pin position with the best TTF. The smallest best TTF over

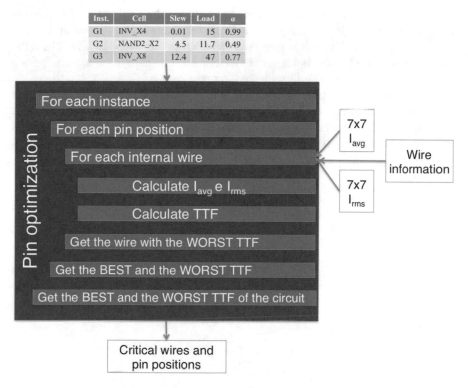

Fig. 5.2 The pin optimization flow used in this work to optimize the lifetime of the circuit avoiding the critical pin positions

the entire circuit is the "weakest link" using the best possible pin positions, and is reported as the **best TTF** of the circuit.

Next, we turn to the problem of optimization, and the objective of our method is to optimize the lifetime of the circuit. We choose the lifetime specification to the best TTF in the circuit. We report the critical pin positions (pin candidates for which the lifetime is smaller than the best TTF) for each cell instance in the circuit, and invalidate these pins. We also enforce a design requirement that limits the maximum allowable Joule heating in a wire. A typical Joule heating specification is a 5 °C temperature rise. We invalidate pin candidates in a cell that violate these requirements, Joule heating, and lifetime. In this way we are trying to increase the number of pin positions for the router, helping to improve the routability, differently if we just selecting the **best TTF**. Our pin optimization tool also can report the TTF for each pin position of each instance in the circuit. This information can be used with other tools to optimize the EM TTF of the circuits.

For the output pin position optimization, we provide the above information to the router, describing pin positions to be avoided. We implement this by changing the pin information in the Library Exchange Format (LEF) file to outlaw the critical pin positions as we build a new TTF-optimized layout, as the last steps in Fig. 5.1

present. The layout of the cells are not changed, just the LEF file for each cell is changed. For the Nangate library used in this work, there is one version of the LEF file for each cell. The circuit is re-routed considering this new LEF file. Then, the timing, power, area, and wirelength are reported to compare if the circuits have a significant change in the results with the pin optimization.

For the Vdd and Vss pins, the LEF file was not changed to avoid the critical pin positions and the circuit was not re-synthesized considering the restricted Vdd and Vss pin positions, because the impact on the global power grid is negligible due to these minor changes. Therefore, it is enough for our analysis to just perform local analyses.

Chapter 6
Results

Tables 6.1, 6.3, and 6.4 present the results of our lifetime evaluation scheme for the set of library cells considering the output, Vdd, and Vss pin placement, respectively, at 2 GHz. The best and worst TTF values correspond to the largest and smallest lifetimes over all pin candidates. The TTF is calculated for two different switching activities (α) of 50 and 100 % of the clock frequency: although few cells in a layout switch frequently, it is likely one of these cells that could be an EM bottleneck. The 100 % switching case is a clear upper bound on the lifetime of the cell: typical cells, even worst-case cells, switch at a significantly lower rate, except on always-on networks such as core elements of the clock network. The tables show that the pin position is important: choosing a good pin position could better balance current flow and improve EM lifetime. It can be noted that the worst TTFs for the X16 cells are extremely small: this is due to the large number of pin choices for such cells, and due to the effects of large currents associated with specific pin positions, as well as divergence effects.

Table 6.1 also shows the TTF results for a 100 % (α) considering a recovery factor of 0.6 and 0.8. *All the other results presented in this work are calculated for a 0.7 recovery factor*, as already cited. The recovery factor is used when the current flows are in opposite directions, as Eq. (3.4) shows. The TTF values in bold are those ones that change when the recovery factor changes, i.e., the currents flowing through the critical edges are in the same direction and the average current is calculated as Eq. (3.5). Changing the recovery factor, from 0.7 to 0.6 or 0.8 the TTF changes from 2 to 17.34 %. The difference is because as larger is the current to be recovered larger is the difference from the previous recovery to the new recovery value. Figure 6.1 presents how much (in %) the TTF reduces when the recovery factor is changed to from 0.7 to 0.6 and how much the TTF increases (negative % values) for a recovery factor of 0.8.

© Springer International Publishing AG 2017
G. Posser et al., *Electromigration Inside Logic Cells*,
DOI 10.1007/978-3-319-48899-8_6

Table 6.1 TTF in years for each cell in the library for the output pin positions

Cell	Recovery = 0.7				Recovery = 0.6		Recovery = 0.8	
	50 % (α)		100 % (α)		100 % (α)		100 % (α)	
	Best TTF	Worst TTF	Best TTF	Worst TTF	Best TTF	Worst TTF	Best TTF	Worst TTF
NAND2_X2	22.03	21.85	10.95	10.85	10.95	10.85	10.95	10.85
NAND2_X4	27.65	20.37	8.75	8.08	8.75	8.08	8.75	8.08
NOR2_X2	24.33	24.30	12.11	12.07	12.11	12.07	12.11	12.07
NOR2_X4	29.61	25.71	14.74	10.75	14.74	**10.41**	14.74	**11.65**
AOI21_X2	28.32	28.30	14.12	14.11	14.12	14.11	14.12	14.11
AOI21_X4	13.13	13.10	6.47	6.43	6.47	6.43	6.47	6.43
INV_X4	23.23	9.90	11.49	4.73	11.49	**4.49**	11.49	**4.99**
INV_X8	33.80	16.92	9.09	1.90	**8.76**	**1.68**	**9.44**	**2.20**
INV_X16	30.80	2.42	15.53	0.20	**13.56**	**0.19**	**18.19**	**0.20**
BUF_X4	25.85	12.93	12.64	6.35	12.64	6.35	12.64	6.35
BUF_X8	40.93	13.55	10.92	1.96	**9.22**	**1.72**	**11.95**	**2.28**
BUF_X16	35.91	3.17	17.65	0.50	**15.38**	**0.49**	**20.71**	**0.52**

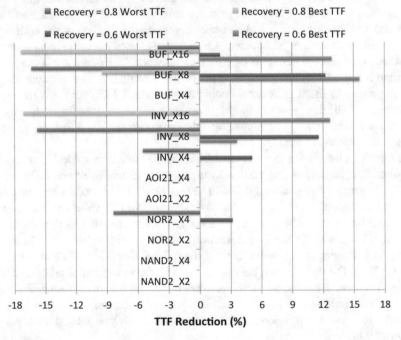

Fig. 6.1 TTF reduction when the recovery factor is changed from 0.7 to 0.6 and 0.8 for a switching activity of 100 % considering the values presented in Table 6.1. Negative values imply in a TTF increase

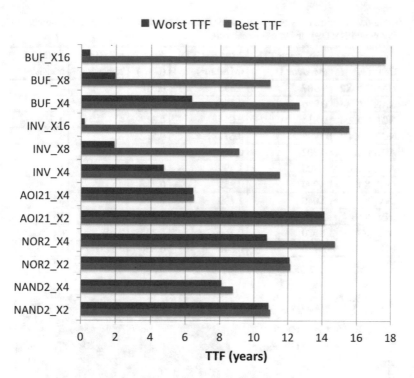

Fig. 6.2 TTF for each cell for different output pin positions for a 100 % switching activity presented in Table 6.1

Figure 6.2 shows a chart representation for the TTF in years for each cell considering the different output pin positions, at 100 % switching activity and a recovery factor of 0.7. The values are the same presented in Table 6.1. In blue are the best TTF that each cell can achieve by placing the output pin in the best position that will have less EM effects.

Table 6.2 shows the current density (J) values for the worst and best TTF presented in Table 6.1 for a 100 % switching activity and a 0.7 recovery factor for each cell. Area (in μm^2) and the number of PMOS and NMOS transistors are also presented, where for the BUF cells the values between brackets are the number of transistors supplying the output. Figure 6.3 shows the current density (J) values presented in Table 6.2, where 1 MA/cm^2 is considered the J boundary, because the EM effects are visible from this current density (ITRS 2011). Remembering that the EM effects are not just influenced by J, they are also influenced by Joule heating that is depended on the RMS current. The J values higher than 1 MA/cm^2 are for the INV, BUF, and AOI21_X4 cells. For the cells with two inputs, J is smaller than the boundary limit and the TTF is not so critical, it is higher or close to 10 years. Just for the NAND2_X4 the worst TTF is 8.08 years.

Placing the Vdd and Vss pins on the best position could improve the INV_X16 lifetime in about 31× and 69×, switching 100 % of the time, as Tables 6.3 and 6.4

Table 6.2 Current density (J) values for the worst and best TTF, area and number of PMOS and NMOS transistors of each cell

Cell	J for worst TTF (MA/cm²)	J for best TTF (MA/cm²)	Area (μm²)	# PMOS transistors	# NMOS transistors
NAND2_X2	0.907	0.898	0.318	4	4
NAND2_X4	0.976	0.652	0.572	8	8
NOR2_X2	0.814	0.814	0.318	4	4
NOR2_X4	0.735	0.671	0.572	8	8
AOI21_X2	0.702	0.701	0.445	6	6
AOI21_X4	1.500	1.490	0.826	12	12
INV_X4	1.990	0.853	0.318	4	4
INV_X8	2.040	1.100	0.572	8	8
INV_X16	2.230	1.230	1.080	16	16
BUF_X4	1.530	0.764	0.445	6 (4)	6 (4)
BUF_X8	2.710	0.846	0.826	12 (8)	12 (8)
BUF_X16	2.810	0.878	1.589	24 (16)	24 (16)

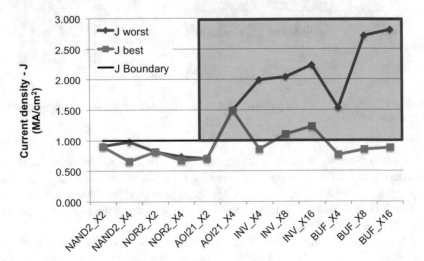

Fig. 6.3 Current density (J) values for the worst and best TTF

show, respectively. These low lifetimes correspond to very high switching rates: in other words, some pin positions would be impermissible on clock buffers, but may be permissible on nets with low switching activity.

For the cell AOI21_X2 the TTF almost doesn't change when the Vdd pin position changes. However, changing the Vss pin position the lifetime can be improved on about 2×. The pin placement has a larger lifetime improvement for the Vss pin than for the Vdd pin. This is because for some cells (AOI21_X2, for example) the geometry of the Vdd and Vss wires are different, producing different pin position options and consequently different current distribution. While this result

Table 6.3 TTF in years for each cell in the library for different Vdd pin positions

Cell	# candidates	50 % switching		100 % switching	
		Best TTF	Worst TTF	Best TTF	Worst TTF
NAND2_X2	6	24.84	22.28	12.38	11.10
NAND2_X4	10	23.66	11.36	11.80	5.57
NOR2_X2	5	51.10	24.13	25.48	12.02
NOR2_X4	6	24.84	12.14	12.39	5.93
AOI21_X2	4	28.34	28.23	14.11	14.05
AOI21_X4	5	26.45	13.40	13.16	6.61
INV_X4	6	18.75	9.03	9.32	4.41
INV_X8	10	18.43	4.31	9.16	1.57
INV_X16	18	15.69	1.42	7.64	0.25
BUF_X4	8	22.45	7.35	11.12	3.52
BUF_X8	14	21.40	3.24	10.37	1.24
BUF_X16	26	11.03	1.24	5.31	0.25
AVG	9.83	–	–	–	–

Table 6.4 TTF in years for each cell in the library for different Vss pin positions

Cell	# candidates	50 % switching		100 % switching	
		Best TTF	Worst TTF	Best TTF	Worst TTF
NAND2_X2	5	41.38	22.57	20.63	11.20
NAND2_X4	5	23.22	10.99	11.52	5.33
NOR2_X2	6	43.05	22.57	21.49	11.20
NOR2_X4	10	43.39	10.81	21.65	4.20
AOI21_X2	6	52.59	25.74	26.26	12.80
AOI21_X4	10	30.56	12.39	14.98	5.39
INV_X4	6	18.67	8.68	9.21	4.10
INV_X8	10	18.35	3.32	9.06	0.95
INV_X16	18	15.68	0.92	7.61	0.11
BUF_X4	8	22.28	7.04	10.93	3.23
BUF_X8	14	21.42	2.77	10.35	0.81
BUF_X16	26	15.68	0.92	7.61	0.11
AVG	10.33	–	–	–	–

may include possible inaccuracies from our direct geometric scaling of the publicly available 45 nm cell layouts to 22 nm, the impact of pin positions is real and can be extreme for large cells. To counter this effect, a cell layout may use wider wires to control current densities, or more practically, outlaw a set of critical positions. For example, for each of the X16 cells, pin positions that see more balanced currents provide high lifetimes (as shown by the best TTF for these cells). More details about the INV_X16 cell are presented in Sect. 6.1.4.

Figure 6.5 shows a chart representation for the TTF in years for each cell considering the different Vdd and Vss pin positions, respectively, at 100 % switching

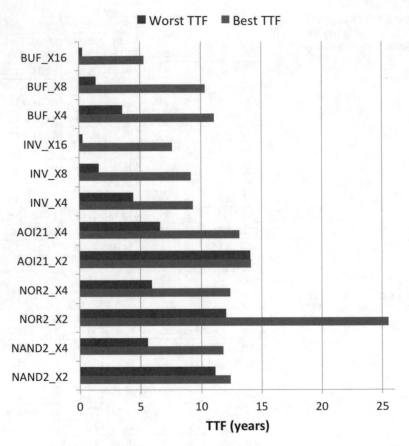

Fig. 6.4 TTF for each cell for different Vdd pin positions for a 100 % switching activity presented in Table 6.3

activity. The values are the same presented in Tables 6.3 and 6.4. In blue are the best TTF that each cell can achieve by placing the Vdd or Vss pin in the best position that will have less EM effects (Figs. 6.4 and 6.5).

The maximum load capacitance is different for each cell. The cells with a larger size can load a larger capacitance without increasing the delay. Considering the load capacitance used in the Nangate cell library, as we are scaling to a lower technology node (22 nm), the max capacitance should usually reduce. This is because the sizes of the loads go down, so the gate capacitance goes down, and also the length of each wire also goes down (distance from driver to load, or to first buffer for a buffered interconnect). So this should be factored into the output load values. Then, we found for each cell the maximum output load that the cell can load without violate the condition $((delay + transition\_time/2) > target\_period)$ and the VDD signal is able to reach the VDD voltage (0.88 V for 22 nm).

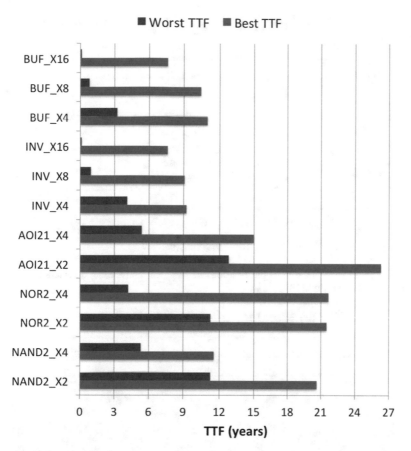

Fig. 6.5 TTF for each cell for different Vss pin positions for a 100 % switching activity presented in Table 6.4

Figure 6.6 shows the TTF in years for the different pin position options for an INV_X4 (Fig. 1.5), considering a switching activity of 100 % at 2 GHz. It is possible to see a lifetime difference between the pin positions. When the pin is at node 4, the TTF is 2× larger than when the pin is at PMOS or at node 2 or node 6 and 2.78× larger than when the pin is at NMOS. For this cell, the best TTF, i.e., the TTF achievable changing the pin position, is 7.38 years and the worst TTF is 2.65 years at 2 GHz.

Figure 6.7 shows the TTF in years for the different output, Vdd, and Vss pin position options for an INV_X4, considering a switching activity of 100 % at 2 GHz. The values for the output pin are the same shown in Fig. 6.6. The different pin positions are named from P1 to P6, where the TTF changes for each different pin position.

In Fig. 6.7, relating to Fig. 1.5, P1 is when the output pin is at node 3, Vdd is at node 2, and Vss is at node 1. P2 is when the output pin is at node 7, Vdd is at node

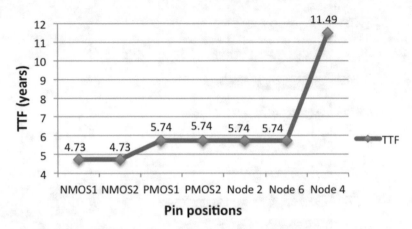

Fig. 6.6 TTF in years for various pin positions in INV_X4, considering a switching activity of 100 % and a frequency of 2 GHz

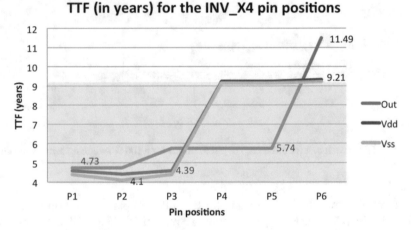

Fig. 6.7 TTF for INV_X4 output, Vdd, and Vss pin positions considering 100 % switching activity

4 and Vss is at node 3. These are the critical pin positions for this cell, where they have the smallest TTF. Avoiding the critical pin positions, the largest TTF that the INV_X4 can achieve is 9.21 years, that is limited by the Vss pin. So, to achieve the maximum TTF, all pin positions with a TTF smaller than 9.21 years are avoided, as the colorful area in the chart shows. The TTF for the INV_X4 can be improved in 2.25× avoiding the critical output, Vdd, and Vss pin positions. For this cell, the best TTF given by the output pin is 11.49 years and this value cannot be achieved because it is limited by the best TTF of the Vss pin, that is 9.21 years. Moreover, the worst TTF of this cell is 4.1 years and it is also given by the Vss pin.

Table 6.5 Timing, power, area, and wirelength reports from the Encounter tool after place-and-route the set of benchmark circuits

Circuit	# of comb. cells	Period (ns)	Power (mW)	Area of core (μm^2)	Total wire length (μm)
b05	859	0.544	0.551	504	2682.50
b07	461	0.306	0.352	317	1426.87
b11	821	0.384	0.460	471	2439.83
b12	1217	0.282	0.810	824	4236.15
b13	340	0.208	0.467	272	1272.99
s5378	1219	0.299	0.679	890	6418.27
s9234	1044	0.373	0.584	849	4873.30
s13207	1401	0.720	1.063	1733	7146.48
s38417	10,068	0.493	8.836	7959	46,419.93
aes_core	27,420	0.345	25.393	13,356	206,199.45
wb_conmax	34,562	0.438	14.228	18,176	321,431.88
des_perf	90,112	0.441	121.190	59,206	727,368.54
vga_lcd	103,774	0.331	70.128	73,450	1,189,099.87

Table 6.6 Cell-internal EM analysis for a set of benchmark circuits computing the best and worst TTF values as described in Chap. 5 for the output pin position

Circuit	Worst TTF (years)	Best TTF (years)	TTF improv.	# of critical nets	# of critical instances	$\frac{critical\_instances}{total\ instances}$ (%)
b05	4.07	6.53	1.60×	—	4	0.47
b07	3.81	5.25	1.38×	—	3	0.65
b11	2.75	5.82	2.12×	1	5	0.61
b12	3.13	3.14	1.001×	3	1	0.08
b13	3.89	6.05	1.56×	1	7	2.06
s5378	2.74	3.59	1.31×	2	1	0.08
s9234	2.73	3.48	1.27×	—	1	0.10
s13207	4.94	13.18	2.67×	—	7	0.50
s38417	3.43	5.77	2.68×	2	6	0.06
aes_core	2.28	5.06	2.22×	63	5	0.02
wb_conmax	2.26	5.25	2.32×	6	59	0.17
des_perf	1.91	5.05	2.65×	10	12	0.01
vga_lcd	0.18	2.87	15.77×	2308	183	0.18

Table 6.5 presents the results for a set of ITC'99 and ISCAS'89 benchmarks circuits mapped to our set of characterized cells and placed-and-routed. For each benchmark the number of combinational cells, the clock period, total power consumption (leakage and switching power), area of core, and total wirelength (WL) are presented, as reported by Encounter.

The best and worst TTF values are computed as described in Chap. 5 and are presented in Table 6.6. These results correspond to a post place-and-route layout with no EM awareness, and the gap between the best and worst TTF values indicates how much the lifetime can be improved. The worst TTF is considering that the router chooses the worst pin position, where the TTF of the cell will be the smallest value. The best TTF is achieved when the critical pin positions are avoided. The number of critical nets corresponds to the nets that violate the Joule heating constraint (5 K), and the number of critical cells corresponds to the cells that have pin positions that corresponds to lifetimes below the best TTF. Interestingly, these numbers are both small, implying that large improvements to the lifetime can be obtained through a few small changes to the layout. Note that the best TTF values are in the range required for many modern applications (e.g., mobile devices) with short TTF specs of 3–4 years. Generally the intended TTF for many electronic interconnects in integrated circuits is approximately 10 years (ITRS 2011; Lienig 2013).

Table 6.6 shows that the lifetime of a circuit can be improved by up to 15.77× by altering the output pin position of a few cells. The benchmark where the TTF improvement is small is b12: the critical cell for this circuit is an NOR2_X2 where the worst TTF is 3.13 and the best TTF is 3.14, i.e., changing the output pin position the TTF does not change the lifetime significantly. The largest TTF improvement is for vga_lcd circuit, where the critical cell is an INV_X8 and its worst TTF is 0.18 years and the best TTF is 2.87 years, given by an instance of an INV_X4 cell.

We now redo the routing step to guarantee that the best TTF in Table 6.6 can be met by outlawing all pin positions whose TTF is worse than the best TTF in Table 6.6, or that result in a cell-internal Joule heating violation. Since the best TTF was computed by choosing the best pin position for each cell, and then finding the weakest link by determining the shortest TTF among these cells, a few cells may be forced to use a single pin, but most cells will have the choice of a number of pin positions, and the circuit lifetime will be significantly enhanced. (Note that by the definition of best TTF, each cell is guaranteed to have at least one allowable pin.)

After these new constraints are imposed on the pin positions, the router makes incremental changes to some interconnect routes. Figure 6.8 shows how the router changed the interconnections around an instance of the INV_X4 and NOR2_X2, respectively, when the critical pin positions are avoided.

Figure 6.8a shows the connections to INV_X4 considering the original LEF file, where the H-shape is the output of the cell and the output pin can be placed anywhere in this H-shape. The same happens for the NOR2_X2 cell in Fig. 6.8c, where the output is the largest metal 1 wire (blue) and it is connected to ZN. While, Fig. 6.8b, d shows the connections when the critical pin positions are avoided and the output pin can be placed just in the center of the INV_X4 (blue connection in the center) and for the NOR2_X2 is the most superior on right metal 1. Thereby, the routes are changed to consider the constrained cell.

After these new constraints are imposed on the pin positions, the router makes incremental changes to some interconnect routes. Table 6.7 shows the results after physical synthesis considering the best output pin positions, i.e., for each cell, we disallow EM-unsafe output pin positions. Thus, we see that the circuit lifetime is improved up to 15.77× while keeping the delay, area, and power of the circuit unchanged, and with marginal changes ($\leq 0.15\%$) to the total wirelength (in fact, for

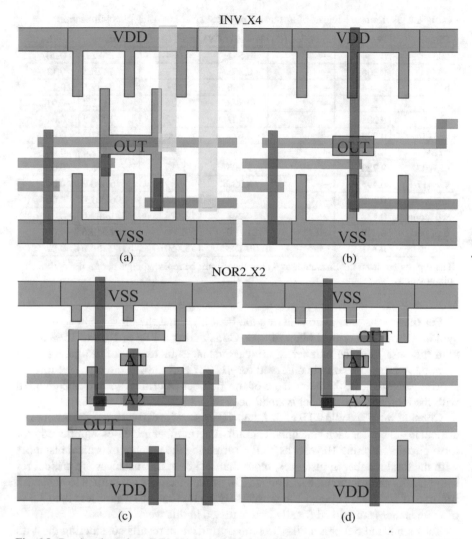

Fig. 6.8 Routing through an INV_X4 (**a**) and (**b**) and an NOR2_X2 cells considering (**a**) and (**c**) the original LEF file and (**b**) and (**d**) avoiding the critical pin positions

two circuits, b12 and des_perf, the wirelength and the clock period are even slightly improved). As there are only a few instances with critical output pin positions and critical wire segments, the TTF can be increased without major changes in the circuit. A set of tests increasing the number of instances that are changed to avoid the EM-unsafe pin positions is presented in Appendix A.

Table 6.7 Performance impact of EM-aware physical synthesis using pin optimization

Circuit	Period (ns)	Δ Period (%)	Power (mW)	Area (μm^2)	WL (μm)	Δ WL (%)
b05	0.544	–	0.551	504	2682.6	0.00
b07	0.306	–	0.353	317	1428.5	0.12
b11	0.384	–	0.460	471	2443.5	0.15
b12	0.280	−0.89	0.808	824	4112.8	−2.91
b13	0.208	–	0.467	272	1273.5	0.04
s5378	0.299	–	0.679	890	6422.2	0.06
s9234	0.373	–	0.584	849	4873.4	0.00
s13207	0.720	–	1.063	1733	7146.6	0.02
s38417	0.493	–	8.836	7959	46,420.2	0.00
aes_core	0.345	–	25.393	13,356	206,207.8	0.00
wb_conmax	0.438	–	14.228	18,176	321,409.6	−0.01
des_perf	0.440	−0.11	121.190	59,206	727,319.6	−0.01
vga_lcd	0.331	–	70.128	73,450	1,189,356.6	0.02

The reports are from the Encounter tool after the circuit be replaced considering the optimized
pin positions

For these tests we are considering the best TTF that the circuit can achieve by
avoiding the output critical pin positions. Considering a TTF 5, 10, and 15 % larger
than the best TTF, the number of critical instances increases a little bit for most
circuits. A larger increase is observed for s13207 and s38417 circuits, as Fig. 6.9a
shows. Nevertheless, the percentage of the number of critical instances compared
with the total instance number is small, less than 3.3 %, as Fig. 6.9b shows.

Considering a specific TTF of 5, 7, and 10 years, Fig. 6.10a shows the number of
critical instances for each benchmark circuit. This number increases significantly for
most circuits targeting 10 years as TTF. Comparing the number of critical instances
with the total number of instances, more than 3.5 % of the instances are critical for
four circuits and for circuit b13, about 21 % of the instances are critical to achieve a
lifetime of 10 years, as Fig. 6.10b presents. For a lifetime of 7 years, less than 3.5 %
of the instances are critical for all circuits tested in this work.

Tables 6.8 and 6.9 present the lifetime optimization results considering the Vdd
and Vss pin placement, respectively. The results are obtained in the same way as
those considering the output pin placement and for the same benchmark circuit
set. The worst and best TTF are shown for each circuit and its TTF improvement.
Furthermore, the number of critical nets and critical cells that have to be avoided
to achieve the best TTF is also shown. For the Vdd pin, avoiding the critical pin
positions the TTF of the circuits can be improved from 1.63× to 81.73×, as shown
in Table 6.8. For most circuits the number of critical cells is very small, about 10. For
the circuits s13207 and s38417 the number of critical cells is 48 and 39, respectively,
representing about 3.4 % of the total number of cells. For the vga_lcd circuit, about
10 % of the instances are critical, i.e., with Vdd pin positions that give a TTF smaller
than 2.94 years.

The results for the Vss pin placement are shown in Table 6.9, where a higher TTF
improvement is possible choosing the best Vss pin position than choosing the best

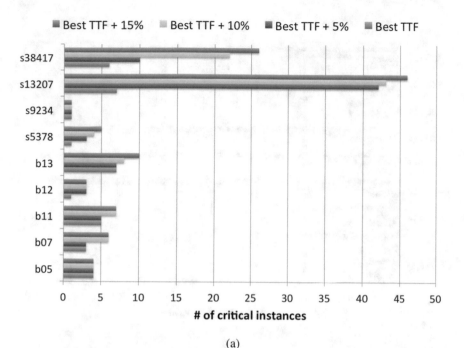

(a)

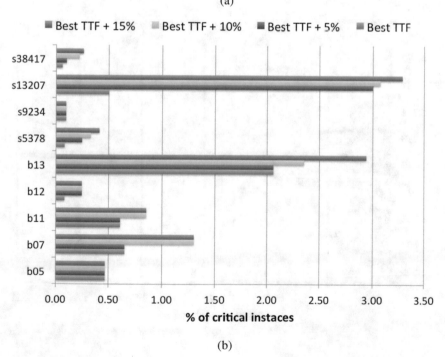

(b)

Fig. 6.9 (**a**) Number (#) and (**b**) percentage (%) of critical instances compared with the circuit total instances increasing the best TTF by 5, 10, and 15 %

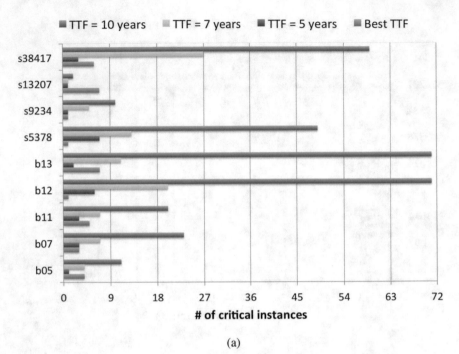

(a)

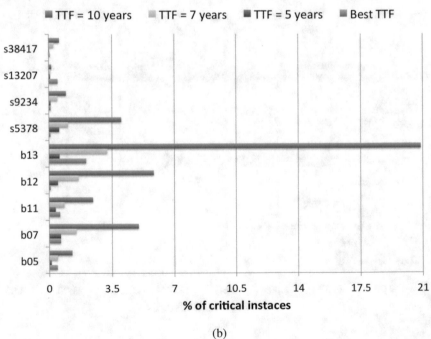

(b)

Fig. 6.10 (**a**) Number (#) and (**b**) percentage (%) of critical instances compared with the circuit total instances for a TTF of 5, 7, and 10 years

Table 6.8 Vdd pin analysis for a set of benchmark circuits

Circuit	Worst TTF (years)	Best TTF (years)	TTF improv.	# of critical nets	# of critical instances	critical_instances / total_instances (%)
b05	4.26	7.87	1.85×	−	7	0.81
b07	1.15	5.36	4.66×	9	7	1.52
b11	2.94	7.18	2.44×	−	10	1.22
b12	2.60	4.23	1.63×	−	8	0.66
b13	2.08	5.06	2.43×	−	9	2.65
s5378	2.40	5.27	2.20×	−	11	0.90
s9234	2.38	6.04	2.54×	−	7	0.67
s13207	5.20	11.85	2.28×	−	48	3.43
s38417	3.73	6.30	1.69×	−	39	0.39
aes_core	0.85	4.66	5.47×	267	64	0.23
wb_conmax	0.37	5.35	14.44×	958	205	0.59
des_perf	1.51	5.84	3.87×	169	274	0.30
vga_lcd	0.04	2.94	81.73×	11,202	1162	1.12

Table 6.9 Vss pin analysis for a set of benchmark circuits

Circuit	Worst TTF (years)	Best TTF (years)	TTF improv.	# of critical nets	# of critical cells	critical_instances / total_instances (%)
b05	3.56	9.37	2.63×	1	14	1.63
b07	1.00	7.77	7.77×	9	29	6.29
b11	2.17	6.64	3.06×	15	10	1.22
b12	1.32	7.34	5.56×	27	47	3.86
b13	1.04	10.29	9.89×	12	204	60.00
s5378	1.22	5.61	4.60×	15	12	0.98
s9234	2.23	5.73	2.57×	8	8	0.77
s13207	4.41	11.31	2.56×	2	12	0.86
s38417	2.51	8.29	3.30×	22	68	0.68
aes_core	0.59	4.35	7.34×	532	68	0.25
wb_conmax	0.25	6.08	24.09×	1041	230	0.67
des_perf	1.31	5.48	4.18×	1067	269	0.30
vga_lcd	0.02	2.51	160.66×	14,164	1238	1.19

output or Vdd pin positions. The TTF can be improved from 2.5× to 161× avoiding the critical Vss pin positions. The number of critical nets and cells is also larger than for output and Vdd pins. The largest number is for b13 circuit, where there are 204 critical cells and this is 60 % of the total number of the cells in the circuit. Des_perf is other circuit with a large number of critical cells, with more than 10 % of the total number of combinational cells. For the other circuits, the number of critical cells is not larger than 3.8 % of the total number of cells of the circuit.

Table 6.10 TTF results optimizing the output, Vdd, and Vss pin positions for a set of benchmark circuits

Circuit	Worst TTF (years)	Best TTF (years)	TTF improv.	# of critical nets	# of critical cells	$\dfrac{\text{critical\_instances}}{\text{total\_instances}}$ (%)
b05	3.56	6.53	1.83×	1	4	0.47
b07	1.00	5.25	5.25×	18	7	1.52
b11	2.17	5.82	2.68×	16	10	1.22
b12	1.32	3.14	2.38×	30	8	0.66
b13	1.04	5.06	4.87×	13	8	2.35
s5378	1.22	3.59	2.94×	17	7	0.57
s9234	2.23	3.48	1.56×	8	3	0.29
s13207	4.41	11.31	2.56×	2	48	3.43
s38417	2.51	5.77	2.30×	24	26	0.26
aes_core	0.59	4.35	7.34×	532	68	0.25
wb_conmax	0.25	5.25	20.83×	1041	220	0.64
des_perf	1.31	5.05	3.86×	1067	263	0.29
vga_lcd	0.02	2.51	160.66×	14,164	1238	1.19

Tables 6.6, 6.8, and 6.9 show the TTF improvement when the output, Vdd, or Vss pin positions, respectively, are optimized separately. In this way, the results when the benchmark circuits are optimized to avoid the critical pin positions simultaneously are shown in Table 6.10. The best TTF of the circuit is the smallest best TTF among the output, Vdd, and Vss pin optimization values. Consequently, the worst TTF is the smallest TTF among the worst TTF of the pin positions. The number of critical cells is reduced compared to the Vss pin optimization because the TTF limit (best TTF) is smaller, reducing the number of critical pin positions and consequently the number of cells. By optimizing the pin positions, the lifetime of the circuits could be improved about 2.5×–161×, that is the case of the vga_lcd circuit, where the lifetime can be improved from 0.02 to 2.51 years, avoiding the critical pin positions. Note that our objective in this work is to obtain the best possible TTF by merely moving pin positions.

For a TTF 5, 10, and 15 % larger than the best TTF achieved when the output, Vdd, and Vss pins are optimized simultaneously, the number of critical instances increases more significantly for the s38417, s13207, and b13 circuits, as Fig. 6.11a shows. Comparing the number of critical instances with the total instance number, just two circuits have more than 1.75 % of critical instances, s13207 and b13. The circuit with the largest percentage of critical cells is b13, with about 8.25 %, as Fig. 6.11b shows.

Considering a specific TTF of 5, 7, and 10 years, Fig. 6.12a shows the number of critical instances for each benchmark circuit. This number increases significantly for most circuits targeting a lifetime of 10 years. For a lifetime smaller than 10 years, all circuits have less than 60 critical instances. Looking the percentage of critical instances compared with the total number of instances, the circuit b13 has

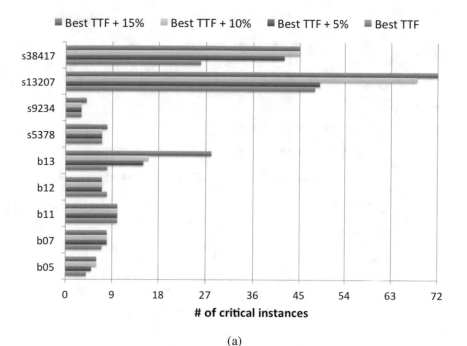

(a)

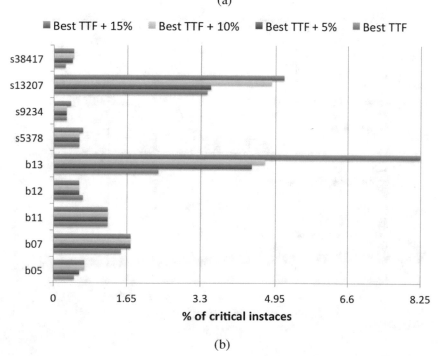

(b)

Fig. 6.11 (**a**) Number (#) and (**b**) percentage (%) of critical instances compared with the circuit total instances optimizing the output, Vdd, and Vss pin positions to achieve an increased best TTF by 5, 10, and 15 %

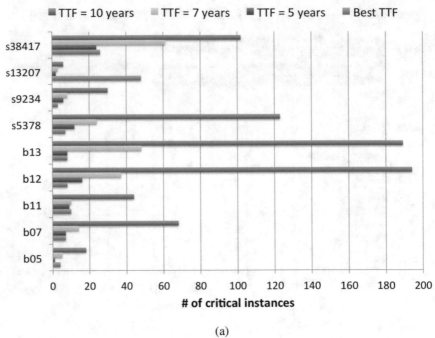

(a)

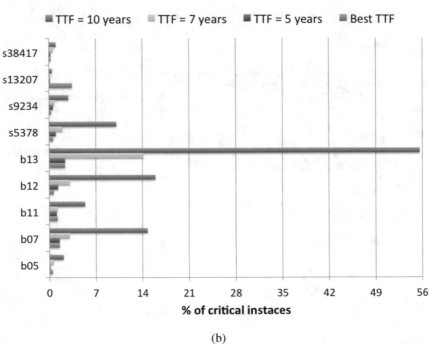

(b)

Fig. 6.12 (a) Number (#) and (b) percentage (%) of critical instances compared with the circuit total instances optimizing the output, Vdd, and Vss pin positions for a TTF of 5, 7, and 10 years

a considerable % of critical instances when the lifetime is larger than 7 years, achieving about 55 % of the instances for a TTF of 10 years, as Fig. 6.12b presents. For other three circuits, s5378, b12, and b07, more than 6 % of the instances are critical for a lifetime of 10 years.

Runtime As previously cited, the circuit analysis is executed by Encounter tool and the runtime for each benchmark is less than 50 s for the benchmarks with up to 10k cells. For the benchmarks aes_core and wb_conmax, the runtime is less than 2 h and for des_perf and vga_lcd, this runtime is about 5–7 h. The critical pin positions for each circuit are reported in under 1 s.

6.1 The Electromigration Effects for Different Logic Gates

Taking into account the cell-internal signal EM analysis for the output signal wire presented in this work, where the current density varies with layout geometry and is a dominant factor in determining the electromigration, this section presents:

1. an analysis of the EM effects on the cell-internal signal wires observing how the current flows through the output wire segments of different logic gates and different output wire geometry. Some examples of the cells are presented to explain why some cells have a larger TTF improvement when the output pin position is changed;
2. some suggestions to optimize the cell layouts to make the cells more robust from the EM perspective.

The idea is to use these robust cells replacing the critical cells that are affected by EM in the circuits. The objective is to increase the circuit TTF beyond the TTF achieved just by the pin position optimization. The critical cells are that ones with the highest switching activity. As shown in Table 6.6, the number of critical cells to be replaced is small, for a 10,000 gates design, is just 6. Thus, the increasing in area or the lost in performance doing the critical cells more robust to EM practically does not affect the overall results.

This study is motivated by the results presented in Table 6.1 showing that the TTF of the cells can be maximized optimizing the output pin placement. We can see that the TTF for some cells almost does not change when the pin position changes, as for the NAND2_X2, NAND2_X4, NOR2_X2, AOI21_X2, and AOI21_X4. On the other hand, the TTF for the other cells can be improved up to 76× (INV_X16) choosing the best pin position. Thus, in the next sections, we are presenting the analysis for some of those logic gates shown in Table 6.1. The simulations are based on scaling the layouts in the NANGATE 45 nm cell library down to 22 nm. The currents through the edges of the cell output signal wire are calculated by SPICE simulation using the publicly available 22 nm PTM model (PTM HP) (Zhao and Cao 2007). The input slew and output load used in the simulations are tested to allow that the circuit operates correctly at 2 GHz, where all transitions can be completed.

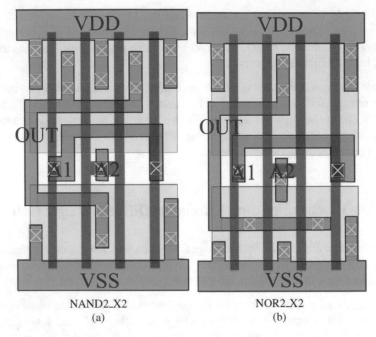

Fig. 6.13 The layout and output pin position options for (**a**) an NAND2_X2 and for an NOR2_X2 (**b**) gate

6.1.1 NAND2_X2 and NOR2_X2 Gates

Figure 6.13a, b shows the layouts of the NAND2_X2 and NOR2_X2 gates. These gates have two inputs: A1 (first and fourth transistors) and A2 (second and third transistors). The output signal for the NAND2_X2 is represented by the 1–8 nodes and for the NOR2_X2 is represented by the 1–6 nodes. These two cells have a very small TTF difference among the different pin position options where the best and the worst TTF are very similar, as Table 6.1 shows.

Figure 6.14 shows the average current values through the edges e_1-e_7 of the NAND2-_X2 output signal wire considering an input slew of 78 ps and an output load of 4.5 fF. As the wire width and wire thickness is the same for all wire segments, the larger current density through the wire segments (edges) will be determined by the larger I_{avg} value among the edges.

The current injection points whose currents contribute to the current in the edges for the rise transition are the nodes 3 and 7 and for the fall transition is just the node 6. Nodes 3 and 7 inject a current of 11.1 μA each one, so the maximum rise current flowing through the edges will be about the sum of these two currents, as Fig. 6.14a, c shows for the edge e_2 (21.4 μA). For the fall transition, the node 6 injects 26.3 μA. If the output pin is at node 1 (Fig. 6.14a, b), the critical edge is e_4

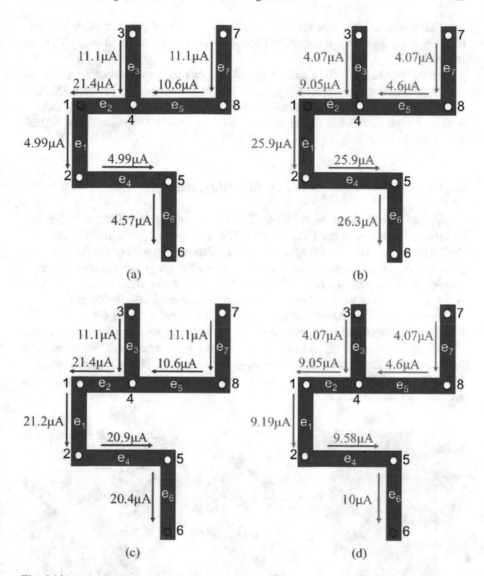

Fig. 6.14 Average current values through the output wire segments of the NAND2_X2 gate when the output pin is at node 1, (**a**) and falling (**b**) and at node 6, (**c**) and (**d**). The *red* [*green*] *lines* represent rise [fall] currents

with an I_{avg} [Eq. (3.5)] of 15.45 μA from node 6. When the output pin is at node 6 (Fig. 6.14c, d), the edge e_4 is also the critical edge with an I_{avg} of 15.24 μA, a little bit smaller current, where the TTF will be not so much different than the TTF when the output pin is at node 1. Placing the output pin at other nodes (2, 3, 4, 5, 7, and 8) the larger current will be very similar to the current presented when the output pin is at node 1. Thus, the best output pin position for the NAND2_X2 is at node 6, where the TTF is a little bit higher than the TTF when the output pin is at other position, as Table 6.1 also shows the small difference between the best and worst TTF for this cell.

For the NOR2_X2 cell, a similar effect happens but in a reverse way. For the NOR2_X2, there is just one node injecting the rise current (node 3 in Fig. 6.13b) and two nodes injecting the fall current (nodes 5 and 6 in Fig. 6.13b). The modifications applied for the PMOS transistors in an NAND2_X2 will be applied for the NMOS transistors in an NOR2_X2 gate.

6.1.1.1 TTF Improvement by Layout Modifications

The output wire shape of the NAND2_X2 gate, as well as the NOR2_X2 gate, can be changed to improve the TTF. Figure 6.15a–c presents three options to improve the lifetime of the cells by changing the shape of the output wire of the NAND2_X2 gate. For the proposed solutions, the most symmetric shape for the output wire is considered. This strategy, in addition with a wider wire segment between nodes 3 and 4, can provide a more balanced current distribution through the output pin.

The first option is presenting a solution with a metal 1 finger for each PMOS transistor joining these currents at node 3. This change will avoid a large current through the edge e_2, as Fig. 6.14a shows, and the larger current will be between the nodes 3 and 4. Thus, the width of the edge between these nodes is increased from 34 to 49 nm. Moreover, the pin should be placed at this edge. This will provide a lifetime increase in 43 %.

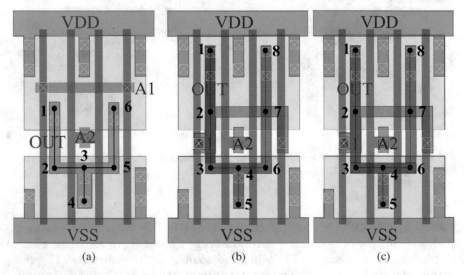

Fig. 6.15 NAND2_X2 layout modification using just metal 1 (**a**), using just metal 2 (**b**) and using metal 1 and 2 (**c**) for the new output wire shape

The second option, presented in Fig. 6.15b, is using just metal 2 for the output wire facilitating the routing. The wire shapes in the second and third options are practically the same used in the first option, where the lifetime of the second option is improved by increasing the width of the edge between nodes 4 and 5 for the second option and of the edges between nodes 3 and 6 for the third option. As metal 2 has a higher thickness than metal 1 and some different properties, its lifetime is larger than metal 1 (Posser et al. 2014c).

The third option, Fig. 6.15c, is using metal 2 for the vertical connections and metal 1 for the horizontal connections of the output wire. The lifetime is improved by increasing the width of the edges between nodes 3 and 6. For the newest technologies, mainly below 32 nm, wider metal in critical nets, and even in some cases the outputs of the cells will use metal 2 to reduce the EM effects (Yeric et al. 2013).

6.1.2 AOI21_X2

The layout of the AOI21_X2 gate (AND, OR, inverter cell) is shown in Fig. 6.16. It has three inputs: A (first and second transistors), B1 (fourth and fifth transistors), and B2 (third and sixth transistors). The output signal is represented by the 1–8 nodes in the center of the cell.

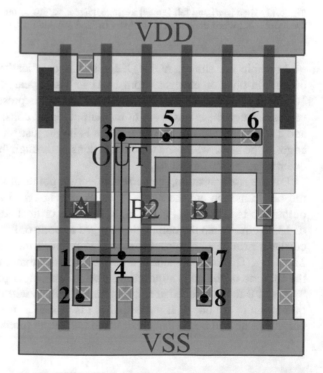

Fig. 6.16 The layout and output pin position options for an AOI21_X2 gate

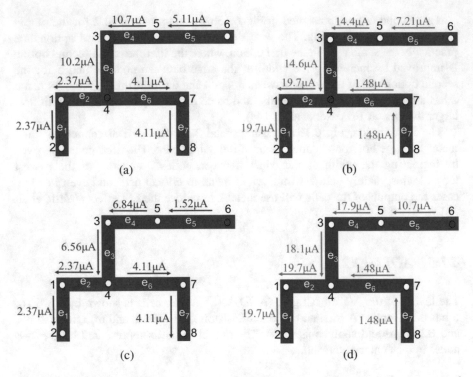

Fig. 6.17 Rise [*red*] and fall [*green*] average current values through the output wire segments of the AOI21_X2 when the output is at node 4 (**a**) and (**b**) and node 6 (**c**) and (**d**)

As Table 6.1 shows, AOI21_X2 gate has practically the same TTF when the output pin position changes (from 14.11 to 14.12 years for a switching activity of 100 %). To understand why this happen, Fig. 6.17 is presenting the average current values through the edges e_1–e_7 of the output signal considering 78 ps as input slew and 1 fF as output load. The wire width is larger just for the edge e_1, for the other edges is the same width and the current density through these wire segments will be determined by the largest I_{avg}.

For the rise transition, the current injection points are the nodes 5 and 6 and for the fall transition are the nodes 2 and 8, as Figs. 6.16 and 6.17 show. When the output pin is at node 4, Fig. 6.17a, b, e_4 is the critical edge with an I_{avg} [Eq. (3.5)] of 12.55 μA. When the output pin changes to node 6, Fig. 6.17c, d, the rise and fall currents through e_4 change, but this edge remains the critical edge with an I_{avg} of 12.37 μA. This current value is a little bit smaller than the previous value, keeping the lifetime of the edge almost the same. If the output pin is placed at nodes 1, 2, 3, 7, and 8 the critical edge and the current values will be the same that when the output pin is at node 4. If the output pin is placed at node 5, the critical edge and current value will be the same that when the output pin is at node 6.

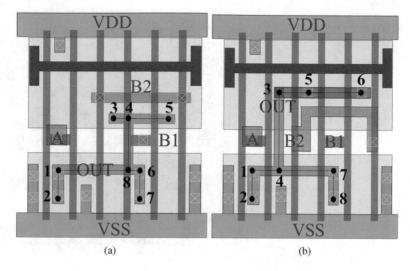

Fig. 6.18 Layout options for an AOI21_X2 gate to improve the TTF

6.1.2.1 TTF Improvement by Layout Modifications

Two different layout options for the AOI21_X2 gate are presented in Fig. 6.18a, b. The lifetime of the AOI21_X2 gate can be improved by changing the shape of the output wire.

In Fig. 6.18a the output pin shape is changed aiming a more symmetric current distribution using just metal 1 and avoiding the large current through the edge e_4 in Fig. 6.17, and the output pin has to be placed at edge between nodes 4 and 8. Figure 6.18b shows a layout improvement where the wire segments of the output from node 3 to node 5 and from node 5 to node 6 are using metal 2, reducing the current density because the metal 2 has a larger thickness, different proprieties and provides a possibility to use a wider wire. The output pin has to be placed at these edges in metal 2.

6.1.3 NOR2_X4

The layout of the NOR2_X4 gate is shown in Fig. 6.19.

This is an NOR gate with size 4, i.e., the transistor widths are four times larger than the minimum width. This gate has two inputs: A1 (second, third, sixth, and seventh transistors) and A2 (first, fourth, fifth, and seventh transistors). The output signal (OUT) is represented by the 1–8 nodes in the center of the cell. As Table 6.1 shows, the TTF of the NOR2_X4 gate can be improved about 37 % when the output pin position changes avoiding the critical pin positions, considering a 100 % switching rate.

Fig. 6.19 The layout and output pin position options for an NOR2_X4 gate

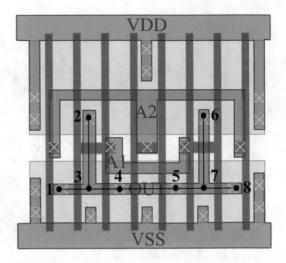

To explain how the current density and consequently the TTF of the cell changes for different pin positions, Fig. 6.20 is presenting the average current values through the edges e_1–e_7 of the output signal obtained by SPICE simulation. For this simulation, we are considering 78 ps as input slew and 8 fF as output load. The wire thickness is the same for all wire segments, and the wire width is 20 % larger just for the edge e_2, for the other edges the wire width is the same. Consequently, the current density through these wire segments will be determined by the largest I_{avg}.

For the rise transition, the current injection points are the nodes 2 and 6 and for the fall transition the current injection points are the nodes 1, 4, 5, and 8, as Figs. 6.19 and 6.20 show. When the output pin is at node 4, Fig. 6.20a, b, e_6 is the critical edge with an I_{avg} [Eq. (3.5)] of 13.18 μA. The same happens when the output pin is at nodes 1, 3, 5, 7, and 8, i.e., the critical edge is e_6. Changing the output pin to node 2 (Fig. 6.20c, d), even that e_2 is wider than the other edges, this is the critical edge with an I_{avg} [Eq. (3.4)] of 14.52 μA. When the output pin is changed to node 6, Fig. 6.20e, f, the critical edge will be e_6 with an I_{avg} of 14.48 μA. This pin position produces the largest current through an edge, so this is the worst pin position for this cell. To find the best TTF (14.74 years, as Table 6.1 shows), the output pin can be placed at someone of these nodes 1, 3, 4, 5, 7, and 8. If the desired TTF is larger than that one achieved by the pin placement, some layout improvements can be made. One example is to increase the width of the edge e_6.

6.1.4 INV_X16

The largest TTF improvements choosing the best output pin position for the cells is for the inverter and buffer gates, where the largest improvement is for the INV_X16 cell, as Table 6.1 shows.

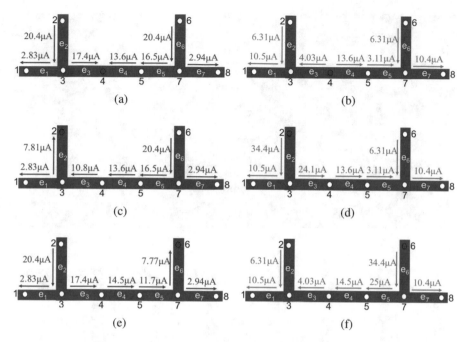

Fig. 6.20 Rise [*red*] and fall [*green*] average current values through the NOR2_X4 output wire segments when the output pin is at node 4 (**a**) and (**b**), node 2 (**c**) and (**d**) and, node 6 (**e**) and (**f**)

Figure 6.21 shows the layout of the INV_X16 (inverter with size 16). The output signal is represented by the 1–25 nodes in the center of the cell. The INV_X16 gate has 16 PMOS injection points and 16 NMOS injection points, as Fig. 6.21a shows. Thereby, depending where the pin is placed the current injected by all PMOS or NMOS injection points is flowing through the same edge, as the example in Fig. 6.21b, where the pin is at node 1. In this case, there is a very large current density through edge e_1 and consequently a very low lifetime compared with other pin positions where the current given by the injection points is better distributed through the edges, that is the case in Fig. 6.21c where the output pin is at node 13, in the middle of the output wire. The horizontal edges (e_3, e_6, e_9, e_{12}, e_{13}, e_{16}, e_{19}, and e_{22}) are two times wider than the vertical edges.

The current injected at each PMOS injection point (1, 4, 7, 10, 14, 17, 20, and 23) is about 8.15 μA and at each NMOS injection point (3, 6, 9, 12, 16, 19, 22, and 25) is about the same, 8.14 μA, as Fig. 6.21c shows. Pin at node 13 is the best pin position for the INV_X16, where the critical edges are the edges connected to the PMOS injection points (e_1, e_4, e_7, e_{10}, e_{14}, e_{17}, e_{20}, and e_{23}). Figure 6.21c also shows that the rise current away node 13 is 26.2 μA from edge e_{12} and the same value from

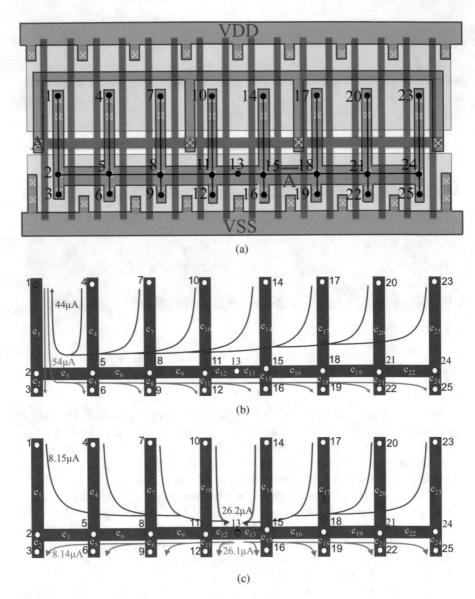

Fig. 6.21 The layout of the INV_X16 (**a**), charge/discharge currents when the output pin is at (**b**) node 1 and (**c**) node 13. The *red* [*green*] *lines* represent rise [fall] currents

edge e_{13}. The fall current through these edges is about the same, but reversed. When the output pin is at node 1, the current from/to all injection points is flowing through e_1, as Fig. 6.21b shows, producing a large current through this edge, reducing its lifetime. This pin position produces the worst TTF for this gate, where the critical edge is e_1.

6.2 Conclusion

In this chapter we presented the results obtained from our cell-internal EM analysis. The results present the importance of the pin placement for the output, Vdd, and Vss pins. The tests executed for the benchmark circuits show that the circuit TTF can be improved up to $15.77\times$ at the same area, delay, and power just avoiding the critical output pin positions. When the output, Vdd, and Vss pin positions are optimized, the lifetime of the circuits could be improved about $2.5\times$–$161\times$.

We are also showing an analysis of the cell-internal signal EM effects for different logic gates considering just the output pin positions. Some logic gates presented a large lifetime improvement possibility changing the output pin position. This is caused by the way that the currents are flowing through the wire segments when the output pin is placed in different positions. On the other hand, for some logic gates the lifetime remains practically unchanged when the output pin position changes. This is because the critical current density, that is through the critical edge, does not change when the output pin position and current flows change. Some layout improvements are suggested to increase the TTF of these cells where changing the output pin position, the TTF practically unchanged.

As a future work, we intend to construct the layouts with the modifications that improve the TTF of the gates. Moreover, the layout parasitics will be extracted and then the gates will be characterized to calculate the accurate TTF of the cells.

Chapter 7
Analyzing the Electromigration Effects on Different Metal Layers and Different Wire Lengths

The analyses, tests, and results presented in the previous chapters of this work are treating the EM effects at cell level, for the wires inside of the cells. In this Chapter we are testing the EM effects at circuit level, on the nets that connect the cells (Posser et al. 2014c). As the nets are signal wires, the direction of current flow is bidirectional and generally is referred as AC electromigration. The AC electromigration has become a serious concern and its limits become tighter with the technology scaling due to the increasing of the on-current of drivers with smaller channel lengths, the decreasing of the interconnect widths, and a faster switching of the transistors increasing the operation frequency (Kahng et al. 2013a). Thus, in this chapter the EM effects on these nets are analyzed for six different metal layers and three different wire lengths, 100, 200, and 300 μm in 22 nm technology. The layouts are constructed considering the 45 nm technology and scaled to 22 nm technology.

The contributions presented in this chapter are as follows:

- To show how the EM affects different metal layers and different wire lengths.
- To analyze the delay behavior for the different metal layers and wire lengths.
- To present how the average current I_{avg} reduces through the wire.

The average and RMS currents are characterized at 2 GHz as a reference frequency, f_{ref}. The simulation results presented in Sect. 6 are considering an activity factor, α, of 100 % at 2 GHz. If the design is operating in a different frequency f and activity factor α, the average and RMS currents can be inferred, they are multiplicatively scaled by factors of $\alpha f / f_{ref}$ and $\sqrt{\alpha f / f_{ref}}$, respectively, as (Posser et al. 2014b) present.

The EM lifetime estimation is computed as presented in Chap. 3, incorporating the Joule heating effects according to Sect. 3.2 shows. As the Joule heating depends on the RMS current I_{rms}, wire resistance R, dielectric thickness t_{ins}, and the thermal conductivity K_{ins}. We obtain I_{rms} by SPICE characterization. R is obtained by parasitic extraction using a commercial tool, t_{ins} changes for different metal layers, as Table 7.1 shows, and $K_{ins} = 0.07$ W/m K (Banerjee and Mehrotra 2001).

© Springer International Publishing AG 2017
G. Posser et al., *Electromigration Inside Logic Cells*,
DOI 10.1007/978-3-319-48899-8_7

Metal layers	45 nm technology (FreePDK45 2011)			22 nm technology		
	W (nm)	T_w (nm)	t_{ins} (nm)	W (nm)	T_w (nm)	t_{ins} (nm)
M1	70	130	120	34	64	59
M2	70	140	120	34	68	59
M3	70	140	120	34	68	59
M4	140	280	290	68	137	142
M5	140	280	290	68	137	142
M6	140	280	290	68	137	142

Table 7.1 Minimum wire width (W), wire thickness (T_w), and the dielectric thickness t_{ins} based on the values from 45 nm technology (FreePDK45 2011) and the values scaled to 22 nm technology

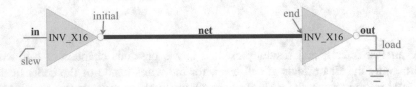

Fig. 7.1 Block diagram used to analyze the characteristics for different metal layers with different wire lengths

7.1 Experimental Setup

Block diagram presented in Fig. 7.1 is the test case used to run the experiments to evaluate the EM effects in different metal layers considering different wire lengths. The block diagram is composed by two INV_X16 from the 45 nm NANGATE Open cell library (Nangate 2011) connected by a net (wire). In the tests, we are changing the metal layer and the wire length of this net. The width of this net is the minimum value given by the technology, as Table 7.1 presents. A different layout is designed for each metal layer and different wire length and then the parasitics are extracted using a commercial tool. The parasitic extractor divides the wire into a number of wire segments. Thus, a separate TTF is calculated for each wire segment and the worst TTF is considered. The worst TTF is the TTF of the closest segment of the first INV_X16, this is because the parasitics of each segment reduce the current flowing through the next segment.

Table 7.1 presents the minimum wire width (W), the wire thickness (T_w), and the dielectric thickness t_{ins} considered in our tests based on the values from the Free PDK 45 nm technology (FreePDK45 2011) and the scaled values to 22 nm technology.

The simulations are executed considering the layout constructed based on the block diagram shown in Fig. 7.1. The structure of the block diagram is kept; just the wire length and the metal layer of the net are changed. The layouts are scaled from the 45 nm Nangate cell library down to 22 nm technology. In 45 nm, the wire lengths used in the layout are 200, 400, and 600 μm. Scaling to 22 nm technology, these wire lengths will be about 100, 200, and 300 μm. SPICE simulation is used to characterize the layout for I_{avg} and I_{rms} values considering 2 GHz as a reference frequency, f_{ref}, and a temperature of 378 K. The simulations are executed considering the scaled 22 nm library based on the 22 nm SPICE ASU PTM model for High Performance applications (PTM HP), and the supply voltage (VDD) used is 0.88 V.

The design is characterized for seven different values each for the input slew and output load, generating a 7×7 look-up table with the RMS and average current values. The input slew values are applied to the *in* signal that is the input of the first inverter in Fig. 7.1. The output load is the capacitance connected to the *out* signal, i.e., to the output of the second INV_X16. The output load of the first INV_X16 is a combination of the net capacitance plus the capacitance of the second INV_X16 plus the output load connected to the second INV_X16. SPICE simulation is used to determine the maximum input slew and output load values. They are limited by the largest values that enable the *out* signal to reach the VDD value and to get an output transition time smaller than the maximum input slew, defined as 198.5 ps. The other constraints are the minimum input slew, defined as 1.2 ps and the minimum output load, equal to 0.6 fF, which is an approximation of the input capacitance of the inverter with size 1 scaled from the 45 nm technology to the 22 nm technology. The TTF values are calculated considering just the current through the wire, the TTF of the vias is not calculated.

7.2 Simulation Results

The results presented in this section are considering the layout of the block diagram shown in Fig. 7.1. Table 7.3 presents the worst TTF for a combination of allowable input slew and output load and the conditions where it occurs, i.e., the input slew and the output load of the worst TTF that are presented in Table 7.2. Moreover, the TTF reduction is also shown for the conditions when the wire length that connects the two INV_X16 in Fig. 7.1 is increased from 100 to 200 μm and from 200 to 300 μm in 22 nm technology. The input slew for the worst TTF was the same for all test cases, and it is the minimum input slew considered in our tests. This is expected because as faster is the input transition, larger is the provided current reducing the TTF. The output load presented in the table is that connected to the second INV_X16 in Fig. 7.1 and its value for the worst TTF is always larger than the constraint 0.6 fF.

About the EM effects, Table 7.3 shows that as lower is the metal layer, lower is the lifetime of the wire. Considering that traditional IC implementation flows have an intended TTF of at least 10 years (Kahng et al. 2013a; Lienig 2013), there are some TTF values in the table smaller than 10 years. And we are considering the test cases with a TTF smaller than 10 years as critical. The wires in metal 1 are all critical, where the criticality is increased as the wire length increases. The wires in metal 2 and metal 3 have a critical TTF for a wire length of 200 and 300 μm, for a wire length of 100 μm the TTF is larger than 10 years. For the metal layers 4, 5, and 6 the TTF is not critical for the wire lengths we are considering in this work, where the smallest TTF for these metal layers is 23.89 years for a wire length of 300 μm.

Table 7.2 Input slew and output load for the different wirelengths of the net in the layout presented in Fig. 7.1

Metal layers	Input slew (ps)			Output load (fF)		
	100 μm	200 μm	300 μm	100 μm	200 μm	300 μm
M1	1.2	1.2	1.2	0.60	0.60	0.60
M2	1.2	1.2	1.2	0.60	0.60	1.20
M3	1.2	1.2	1.2	0.60	0.60	0.60
M4	1.2	1.2	1.2	2.00	0.60	0.60
M5	1.2	1.2	1.2	2.00	0.60	0.60
M6	1.2	1.2	1.2	2.00	0.60	7.50

Table 7.3 TTF and the TTF reduction when the wire (net) length of the layout presented in Fig. 7.1 is changed from 100 to 200 μm and from 200 to 300 μm. Bold values indicate the TTF reduction in percentage when wirelength increases from 100 to 200 μm and from 200 to 300 μm respectively

Metal layers	TTF (years)			TTF reduction (%)	
	100 μm	200 μm	300 μm	$\frac{200\,\mu m}{100\,\mu m}$	$\frac{300\,\mu m}{200\,\mu m}$
M1	8.59	5.71	5.49	**33.53**	**3.85**
M2	11.06	6.94	5.50	**37.25**	**20.75**
M3	13.10	8.54	6.85	**34.81**	**19.79**
M4	52.52	33.00	23.89	**37.17**	**27.61**
M5	58.90	38.88	28.75	**33.99**	**26.05**
M6	63.00	42.82	31.99	**32.03**	**25.29**

The TTF reduction when the wire length is increased from 100 to 200 μm is about 35 % and from 200 to 300 μm the TTF is reduced from 3.85 to 27.61 %.

Accordingly with our tests, TTF is smaller for lower metal layers, even when the metal layers have the same wire width, wire length, and wire thickness. One reason for this is the parasitic capacitance on the wire. Looking the parasitic extraction file, as possible to see that the parasitic capacitances are larger as lower is the metal layer.

Figure 7.2 shows the delay (ps) by metal layers considering the three different wire lengths 100, 200, and 300 μm of the test cases presented in Tables 7.2 and 7.3. The figure shows that the delay increases as the wire length increases because the wire resistance increases. Furthermore, as higher is the metal layer, smaller is the delay because the wire width and wire thickness are higher, reducing the wire resistance.

Table 7.4 presents the maximum input slew and the maximum output transition time (tt) from 5 to 95 % (95–5 %) of the output signal in ps for the different metal layers and wire lengths tested in this work. The maximum input slew we are considering, 198.5 ps, is about 40 % of the clock period (500 ps).

All the maximum output transition time values shown in Table 7.4 are smaller than the maximum input slew. Thus, the second inverter is able to load other cells and wires since the output signal can reach the VDD value and the output transition time is respected. For metals 1, 2, and 3 and wire length of 300 μm, the input slew

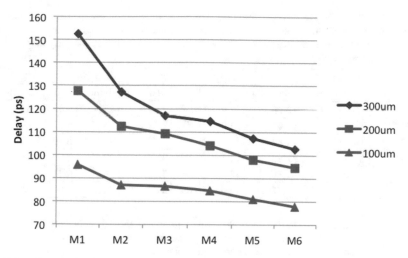

Fig. 7.2 Delay (ps) by metal layers considering the three different wire lengths in 22 nm 100 μm, 200 μm, and 300 μm

Table 7.4 Max input slew and max output transition time (tt) for the different metal layers and wire lengths used in this work

Metal layers	100 μm		200 μm		300 μm	
	Input slew (ps)	Output tt (ps)	Input slew (ps)	Output tt (ps)	Input slew (ps)	Output tt (ps)
M1	198.5	142.4	198.5	155.1	17.2	197.2
M2	198.5	144.0	198.5	155.6	130.0	174.0
M3	198.5	142.6	198.5	156.6	180.0	161.8
M4	198.5	142.4	198.5	162.5	198.5	171.0
M5	198.5	137.9	198.5	160.5	198.5	172.1
M6	198.5	130.6	198.5	160.3	198.5	170.0

has to be smaller than the maximum value because using larger input slew than the values presented in the table the output transition time constraint is not respected.

Table 7.5 presents the effective average current I_{avg} in μm, calculated by Eq. (3.3), for the tests presented in Table 7.2, i.e., for that input slew and output load. Table shows I_{avg} value calculated at the point where the net starts (initial), that is the point connected to the via that is connected to the output of the first INV_X16. And the average current at the point where the net ends (end), that is the point connected to the via that is connected to the input of the second INV_X16 in Fig. 7.1.

Table 7.5 shows that the average current reduces significantly along the wire. For a wire with 100 μm, the current reduces from 1.77 to 2.26× until reaching the via connected to the second inverter. The metal 1 has the largest current reduction. For a wire length of 200 μm, the current is reduced from 2.22 to 3.49× along the wire. For a wire with 300 μm, the current reduces from 3.33 to 4.62× through the wire in our test cases.

Table 7.6 shows that increasing the wire length, the I_{avg} current at the *initial* point reduces from 63.2 to 89 % when the wire length is increased from 100 to 300 μm. At

Table 7.5 The effective average current (I_{avg}) in μA at the point where the net starts (initial) and at the point where the net ends (end) for the different wire lengths. Bold values indicate the TTF reduction in percentage when wirelength increases from 100 to 200 μm and from 200 to 300 μm respectively

Metal layers	100 μm			200 μm			300 μm		
	Initial	End	Red. (×)	Initial	End	Red. (×)	Initial	End	Red. (×)
M1	15.5	6.87	**2.26**	21.9	6.27	**3.49**	25.3	5.48	**4.62**
M2	14.4	6.97	**2.07**	21.0	6.71	**2.22**	25.6	6.11	**4.19**
M3	13.1	6.94	**1.89**	18.7	6.73	**2.78**	22.6	6.20	**3.65**
M4	14.5	7.03	**2.06**	21.1	6.85	**3.08**	27.4	6.63	**4.13**
M5	13.2	7.04	**1.88**	18.7	6.88	**2.72**	24.0	6.71	**3.58**
M6	12.5	7.06	**1.77**	17.4	6.89	2.53	22.3	6.69	3.33

Table 7.6 The I_{avg} reduction (Red.) in × and the I_{avg} reduction at the begin and end points when the wire length is increased considering the values presented in Table 7.5

Metal layers	Initial comparison (%)			End comparison (%)		
	$\frac{200\,\mu m}{100\,\mu m}$	$\frac{300\,\mu m}{200\,\mu m}$	$\frac{300\,\mu m}{100\,\mu m}$	$\frac{200\,\mu m}{100\,\mu m}$	$\frac{300\,\mu m}{200\,\mu m}$	$\frac{300\,\mu m}{100\,\mu m}$
M1	41.3	15.5	63.2	8.7	12.6	20.2
M2	45.8	21.9	77.8	3.7	8.9	12.3
M3	42.8	20.9	72.5	3.0	7.9	10.7
M4	45.5	29.9	89.0	2.6	3.2	5.7
M5	41.7	28.3	81.8	2.3	2.5	4.7
M6	39.2	28.2	78.4	2.4	2.9	5.2

the end *point* of the wire, the I_{avg} current reduces from 4.7 to 20.2 % when increasing the wire length from 100 to 300 μm.

7.3 Conclusion

An analysis of the EM effects on different metal layers for different wire lengths was shown in this chapter. We can conclude that as lower is the metal layer, lower is the lifetime of the wire. Then, higher metal layers have smaller EM effects and consequently a higher lifetime for the wires. We are considering critical the nets with a TTF smaller than 10 years. The wider wires have a larger TTF because the current density through these wires is smaller, reducing the EM effects. The signal nets in metal 1 in out test cases for wire lengths of 100, 200, and 300 μm are critical and the lifetime is reduced as the wire length increases. The wires in metal 2 and metal 3 have a critical TTF for a wire length of 200 and 300 μm. For the metal layers 4, 5, and 6 the TTF is not critical for the wire lengths we are considering in this work. The delay in our test cases increases when the wire length increases and decreases for a higher metal layer, i.e., as lower is the metal layer higher is the delay.

Chapter 8
Conclusions

We have developed an approach to touch upon the problem of cell-internal EM, addressing the problem of EM on signal interconnects and on Vdd and Vss rails within a standard cell. A new modeling approach that includes Joule heating effects and current divergence is presented. Based on the review through the literature, few works are concerned with cell-internal EM. To our knowledge, there are no other published approach addressing this problem directly. Thus, our work is the first one to optimize the pin positions of the output, Vdd, and Vss wires in standard cells improving the circuit lifetime.

The main contributions of this work are summarized as follows:

- **A study of the problem of analyzing the EM effects inside standard cells for the output, Vdd, and Vss wires:** we show that the current that flows through the local interconnect wires can be large enough to create significant EM effects over the lifetime of the chip.
- **Cell-internal EM modeling:** The EM was modeled incorporating Joule heating effects and the current divergence to estimate the lifetime of the signal and supply wires.
- **An approach to efficiently characterize cell-internal EM over all output, Vdd, and Vss pin locations within a cell using a reference pin position:** Wherein a graph-based algorithm is used to compute the currents through each edge when the pin position is moved from the reference case to another location. This algorithm speeds up the characterization by the number of different pin positions along with the AVG and RMS currents computation, producing a small calculation error compared to SPICE simulation, just 0.53 % on average.
- **The pin placement optimization problem formulation for the output, Vdd, and Vss pins, where the lifetime of the overall design is maximized:** The pin optimization reports the critical pin positions to be avoided to maximize the circuit lifetime. They are avoided by just changing the Library Exchange Format (LEF) file of the cells, the cell layouts are not changed. Applying this optimization, the circuit lifetime could be improved up to 15.77× for the

© Springer International Publishing AG 2017

G. Posser et al., *Electromigration Inside Logic Cells*,

DOI 10.1007/978-3-319-48899-8_8

benchmark set used in this work. This improvement was possible by just avoiding the critical output pin positions, keeping the circuit area, delay, power, and wirelength. When the output, Vdd, and Vss pin positions are optimized, the lifetime of the circuits could be improved about 2.5×–161×.

8.1   Future Works

As we have already presented, there are few works concerned with the EM effects through the wires within a cell. We presented that the lifetime of the circuit could be improved substantially avoiding the critical pin positions of the output, Vdd, and Vss wires. Thereby, we intend to continue this work where some possibilities are presented below.

1. The EM problem is intensifying with the latest technologies making it necessary to more robust designs, wider metals in critical nets, and even in some cases the outputs of the cells will use metal 2 to reduce the EM effects (Yeric et al. 2013). Thus, we intend to study the EM effects considering two metal layers on the output signal of the cell, wherein the first metal is used for the vertical interconnections and metal 2 for the horizontal interconnections like the new technologies are using (Ban et al. 2014; Vaidyanathan et al. 2014a), as Fig. 8.1. Moreover, in this case the EM effects on the vias from metal 1 to metal 2 should also be considered.
2. To synthesize the circuits avoiding the critical Vdd and Vss pin positions.

Fig. 8.1 Layout of the INV_X4 where the output pin is using two metal layers, metal 1 for the vertical interconnections and metal 2 for the horizontal interconnections. Source: from author (2015)

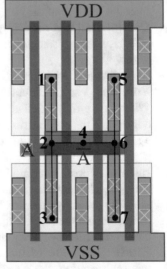

Source: from author (2015).

Fig. 8.2 Layout of the
INV_X4 with the output,
Vdd, Vss pins placed. The
output of the cell is connected
to other two cells. Source:
from author (2015)

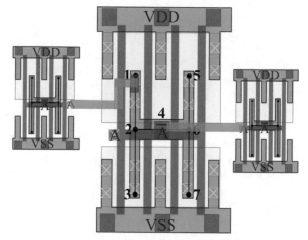

Source: from author (2015).

3. To consider more than 1 pin placed simultaneously on the Vdd and Vss rails. And also on the output wire for cells with a fanout larger than one, i.e., driving more than one cell as Fig. 8.2 shows.
4. To estimate the temperature, used to calculate the TTF, for each cell considering a heat map from a commercial tool. The idea is to use this temperature information to analyze the final TTF, considering the global information to apply to the cells.
5. To study how the commercial tools like Encounter (Cadence 2013) and Virtuoso (Cadence 2015) from Cadence analyze the EM and the current density through the circuit wires. The type of wires that these tools are able to analyze will also be studied and reported.
6. To construct the layouts with the modifications that improve the TTF of the logic gates where the lifetime remains practically unchanged when the output pin position changes or when the lifetime is below of a given specification. One layout modification is to do the critical wires wider to increase its lifetime. Moreover, the layout parasitics will be extracted and then the gates will be characterized to calculate the accurate TTF of the cells.
7. We would also like to test the EM effects considering cell layouts constructed using the 32/28 nm Synopsys PDK library (Synopsys 2014b) and the FreePDK in 15 nm (FreePDK15 2014) used to construct the NanGate FreePDK15 Open Cell Library (Nangate 2014). Furthermore, we intend also use libraries from foundries in recent technologies, like TSMC, Global Foundries.

Appendix A
Impact on Physical Synthesis Considering Different Amounts of Instances with EM Awareness

As Chap. 6 presented, the TTF of the cells can be improved avoiding the critical pin positions that produce high current densities through some wire segments and consequently EM effects in these wires. For this, a pin placement problem was formulated to achieve the best TTF of a circuit placing the output, Vdd, and Vss pins. The physical synthesis was executed for a set of benchmark circuits with no EM awareness and after for the "best" output pin positions (we did not execute yet the physical synthesis avoiding the critical Vdd and Vss pin positions), where the number of critical cells is small. In this way, to consider the constrained pin positions for these cells, the router makes changes to some interconnect routes. As the number of critical cells is small, the results after physical synthesis considering the best pin positions shows a TTF improvement up to 62.50 % while keeping the delay, area, and power of the circuit unchanged.

In this way, this appendix is presenting a set of tests to show the percentage of cell instances that can have EM-unsafe pin positions avoided with a low impact in area, delay, power, and wirelength. The physical synthesis is executed for the set of benchmark circuits for these conditions:

- with no EM awareness, considering the traditional situation with all output pin positions from the original LEF file (original results).
- for the "best" output pin positions, that are the results presented in Chap. 6 (best TTF results).
- considering a TTF of 10 years, where there are a large number of critical cell instances (10 years TTF results).
- changing about 50 % of the total number of instances of the circuit to avoid the EM-unsafe pin positions of these cells (50 % instances changed results).
- avoiding the EM-unsafe pin positions of about 100 % of the instances of the circuit (100 % instances changed results).

© Springer International Publishing AG 2017
G. Posser et al., *Electromigration Inside Logic Cells*,
DOI 10.1007/978-3-319-48899-8

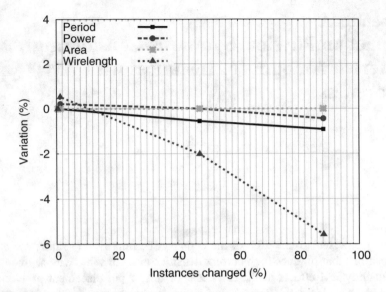

Fig. A.1 Period, power, area, and wirelength variation (%) when the number of instances changed (%) to avoid the critical pin positions increases for the benchmark b05

Then, the area, delay, power, and wirelength are reported for each one of these conditions to show what is the impact when the number of cells that have their output pin positions restricted increases.

Figures A.1, A.2, A.3, A.4, A.5, A.6, A.7, A.8, and A.9 show the period, total power (leakage + switching power), area of core, and the total wirelength variation (%) compared to the original results when the number of instances changed (%) to avoid the critical pin positions increases. The positive values show that the characteristics are improved and the negative values show the opposite. The four points in the charts are representing the following situations:

1. the first point are the results considering the best TTF results compared to the original results.
2. the second point are the results when a TTF of 10 years is considered compared to the original results.
3. the third point are the results when about 50 % of the instances are changed to avoid the critical pin positions compared to the original results.
4. the fourth point are the results when about 100 % of the instances are changed to avoid the critical pin positions compared to the original results.

Figure A.1 shows the variation results for the benchmark b05, where the area is kept the same of the original results. The period and power increase less than 1 % and the wirelength increases about 2 % when about 50 % of the cells have the critical pin positions avoided and 6 % when about 100 % of the instances avoid the EM-unsafe pin positions.

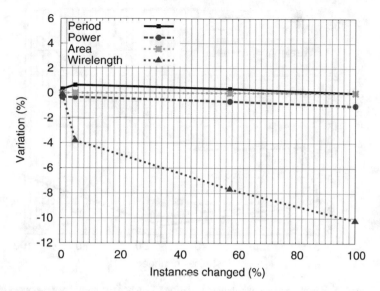

Fig. A.2 Period, power, area, and wirelength variation (%) when the number of instances changed (%) to avoid the critical pin positions increases for the benchmark b07

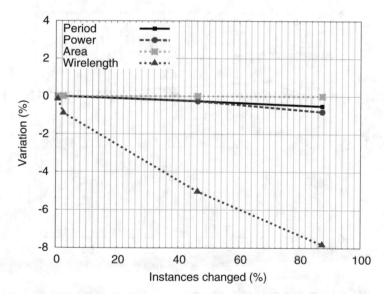

Fig. A.3 Period, power, area, and wirelength variation (%) when the number of instances changed (%) to avoid the critical pin positions increases for the benchmark b11

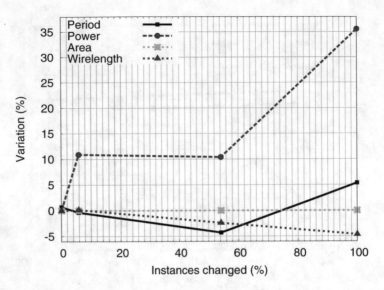

Fig. A.4 Period, power, area, and wirelength variation (%) when the number of instances changed (%) to avoid the critical pin positions increases for the benchmark b12

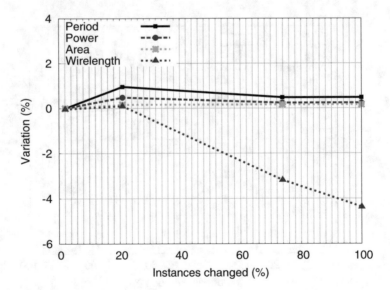

Fig. A.5 Period, power, area, and wirelength variation (%) when the number of instances changed (%) to avoid the critical pin positions increases for the benchmark b13

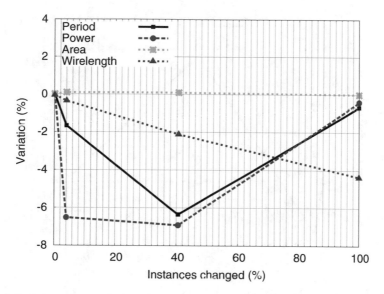

Fig. A.6 Period, power, area, and wirelength variation (%) when the number of instances changed (%) to avoid the critical pin positions increases for the benchmark s5378

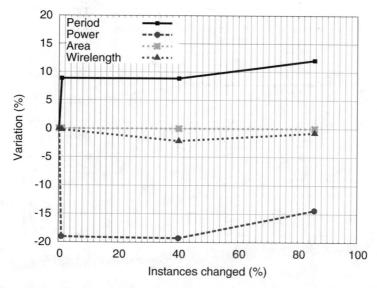

Fig. A.7 Period, power, area, and wirelength variation (%) when the number of instances changed (%) to avoid the critical pin positions increases for the benchmark s9234

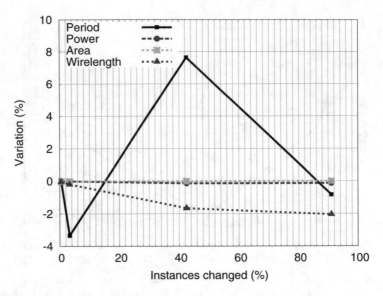

Fig. A.8 Period, power, area, and wirelength variation (%) when the number of instances changed (%) to avoid the critical pin positions increases for the benchmark s13207

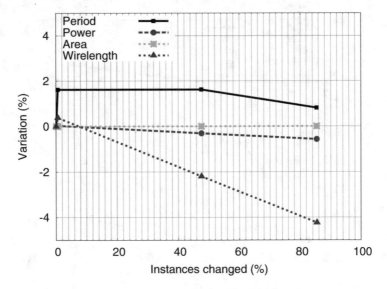

Fig. A.9 Period, power, area, and wirelength variation (%) when the number of instances changed (%) to avoid the critical pin positions increases for the benchmark s38417

The variation results for the benchmark b07 are shown in Fig. A.2. Area is kept the same of the original results. The period and power change less than 1 % and the wirelength increases a maximum of 10 % when 100 % of the instances avoid the EM-unsafe pin positions.

The variation results for the benchmark b11 are shown in Fig. A.3. Area is kept the same of the original results. The period and power reduce less than 1 % and the wirelength increases about 5 % when 50 % of the instances are avoiding the critical pin positions. When 100 % of the instances avoid the EM-unsafe pin positions, the wirelength increases about 8 %.

Figure A.4 shows the variation results for the benchmark b12. For this circuit, the area is kept the same of the original results. The other characteristics have a larger variation than the variation in other circuits. The wire length increases up to 5 % when 100 % of the instances avoid the EM-unsafe pin positions. The period increases about 5 % when about 50 % of the instances are avoiding the critical pin positions and the period reduces more than 5 % when 100 % of the instances avoid the EM-unsafe pin positions. As more instances have the critical pin position avoided, more the power is reduced. The power could be reduced about 35 % avoiding the critical pin positions of the all instances in this circuit. For this circuit, the routes did by the router when the pin positions are restricted for all instances in the circuit were better than the previous routing, with less pin position constraints. Doing the synthesis considering different combinations of the restricted instances, the results can change considerably for this circuit. For example, keeping the power like the power for the original synthesis and reducing the wirelength; reducing the area of core or; just keeping the characteristics close to the original synthesis. The routing is greatly affected by the output pin positions of some cells, where a more constrained routing (due the critical pins avoiding) can produce better results than an unconstrained, where all output pin positions can be used.

The variation results for the benchmark b13 are shown in Fig. A.5. Area is kept practically the same of the original results. The period and power can be reduced about up to 1 %, where the higher reduction is when the critical pin positions are avoided to achieve a TTF of 10 years. The wirelength increases about 3 % when 50 % of the instances are avoiding the critical pin positions. When 100 % of the instances avoid the EM-unsafe pin positions, the wirelength increases less than 5 %.

Figure A.6 shows the variation results for the benchmark s5378. Area is kept the same of the original results. The worst results are presented when 50 % of the instances are avoiding the critical pin positions, where the period increases more than 6 %, power increases about 7 %, and the wirelength increases about 2 %. When 100 % of the instances avoid the EM-unsafe pin positions, the period and power increase less than 1 %, while the wirelength increases about 4 %.

The variation results for the benchmark s9234 are shown in Fig. A.7. Area is kept the same of the original results. The period could be reduced about 10 %. Power increases about 20 % when a TTF of 10 years is considered and when about 50 % of the instances are avoiding the critical pin positions. When about 100 % of the instances avoid the EM-unsafe pin positions, the power increases about 15 %. The wirelength increases less than 1 %.

Figure A.8 shows the variation results for the benchmark s13207. Area and power are kept practically the same of the original results. The period is improved in almost 8 % when 50 % of the instances are avoiding the critical pin positions. For the other cases, it is slightly increased. The wirelength increases about 2 % when more than 40 % of the instances are changed.

The variation results for the benchmark s38417 are shown in Fig. A.9. Area is kept the same of the original results. The period could be reduced about 1.8 %. Power increases less than 1 %. And the wirelength increases about 2 % when about 50 % of the instances are avoiding the critical pin positions. When about 100 % of the instances avoid the EM-unsafe pin positions, the wirelength increases about 4 %.

Appendix B
Coupling Capacitance Currents

The coupling capacitances are considered in our graph-based algorithm (Algorithm 4.1) that computes the currents through each edge when the pin position is moved from the reference case to another location. This Appendix is explaining why the coupling capacitances had to be included to compute accurately the currents using the Algorithm.

Figure B.1 shows the layout of the NAND2_X2 cell, where the Vdd and Vss pins are placed on the middle of these rails. This cell has two inputs, input A1 that are the second and third transistors and input A2, the first and fourth transistors.

The average current direction and values in Ampere (A) through the wire segments of the Vdd rail, when the Vdd pin is at middle (node 3), are presented in Fig. B.2. The red lines and values are the charge current values and the blue lines and values are the short-circuit currents. This example is shown the critical input combination case, where more current is flowing through the edges (causing more EM), A1=0 and A2=1. In this way, the edge from node 3 to node 4 is charging the current through two transistors to the output. This is the critical wire segment when the pin is at middle (node 3).

Figure B.3 shows the average current values when the Vdd pin is placed at node 1. The arrows and values in gray represent the currents that are kept the same than when the pin is at node 3. Just the currents through the edge from node 1 to node 3 change. Based on the values from the reference pin position (node 3), the charge current through this edge is given by $3.89e-5 - 2.32e-8 = 3.888e-6$ and the short-circuit current is calculated by $6.04e-7 + 1.17e-6 = 1.77e-6$, accordingly with the Algorithm 4.1. The calculated values are very similar to the values given by SPICE simulation. The same behavior is seen when the pin is placed at node 5.

Figure B.4 shows the average current values when the Vdd pin is placed at node 2. Figure B.4a, b shows the values from SPICE simulation and (c) and (d) is the currents calculated considering just the current values and directions from the reference pin position. The values from SPICE simulation are different of the values expected considering the reference pin position (node 3), represented by the arrows

© Springer International Publishing AG 2017
G. Posser et al., *Electromigration Inside Logic Cells*,
DOI 10.1007/978-3-319-48899-8

Fig. B.1 Layout of the
NAND2_X2 cell

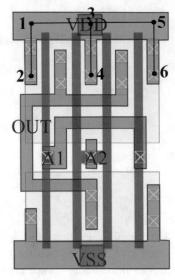

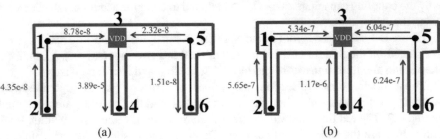

(a)   (b)

Fig. B.2 (**a**) Charge current (in A) and (**b**) short-circuit current (in A) through the wire segments
of the VDD rail when the Vdd pin is at middle (node 3) from SPICE simulation

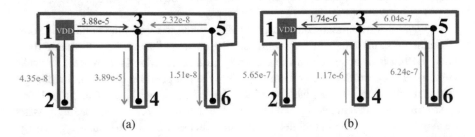

(a)   (b)

Fig. B.3 (**a**) Charge current (in A) and (**b**) short-circuit current (in A) through the wire segments
of the VDD rail when the Vdd pin is at left (node 1) from SPICE simulation

and values in gray, Fig. B.4c, d. For example, the currents through the edge from
node 3 to node 4 should be 3.89e−5 and 1.17e−6 (instead of 3.6e−5 and 1.47e−6,
respectively) accordingly with the values when the pin is at node 3.

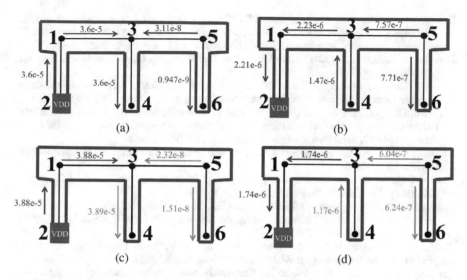

Fig. B.4 (**a**), (**c**) Charge current and (**b**), (**d**) short-circuit current through the wire segments of the VDD rail when the Vdd pin is at node 2. (**a**) and (**b**) values are from SPICE simulation and (**c**) and (**d**) are calculated just considering the currents and directions

For the NAND2_X2 cell, the algorithm just considering the charge and short-circuit current works well when the pin is at node 1, 3, and 5, but do not work well when the pin is at the fingers, nodes 2, 4, and 6. To find why this difference in the values appears for the NAND2_X2 cell, because for other cells like INV and BUF this difference does not happen, we looked through the layout extracted parasitics and we found some coupling capacitances between the VDD and the OUT wires that are producing this difference in the results.

Considering that the coupling capacitance currents are the same for almost all nets since moving the pin does not significantly change the transient waveforms in these nets. Thus, to consider the case of the NAND2_X2 cell, the change is simple: for the reference case, the currents through the wires as well as through the coupling capacitances would be measured. In this way, the charge/discharge, short-circuit/leakage, and coupling capacitance currents for each edge are determined from the simulation to improve the accuracy of our algorithm.

References

Abella J, Vera X, Unsal O, Ergin O, Gonzalez A, Tschanz J (2008) Refueling: preventing wire degradation due to electromigration. IEEE Micro 28(6):37–46
Agarwal KB, Nassif SR, Rose RD, Xu C (2014) Rapid estimation of temperature rise in wires due to joule heating. US Patent 8,640,062
Balhiser D, Gentry J, Harber R, Haskin B, Marcoux P, Stong G (2005) Process and system for identifying wires at risk of electromigration. US Patent App. 10/241,623. http://www.google.com.br/patents/US20040049750

Ban Y, Choi C, Shin H, Kang Y, Paik WH (2014) Analysis and optimization of process-induced electromigration on signal interconnects in 16nm finfet soc (system-on-chip). In: SPIE advanced lithography. International Society for Optics and Photonics, pp 90,530P–90,530P

Banerjee K, Mehrotra A (2001) Global (interconnect) warming. IEEE Circuits Devices Mag 17(5):16–32

Barwin J, Bickford J (2013) Method of managing electro migration in logic designs and design structure thereof. US Patent 8,560,990. http://www.google.com/patents/US8560990

Barwin J, Chung J, Joshi A, Livingstone W, Sigal L, Worth B, Zuchowski P (2015) Identifying and mitigating electromigration failures in signal nets of an integrated circuit chip design. US Patent 9,104,832. https://www.google.com/patents/US9104832

Black JR (1969) Electromigration a brief survey and some recent results. IEEE Trans Electron Devices 16(4):338–347

Blech IA (1976) Electromigration in thin aluminum films on titanium nitride. J Appl Phys 47(4):1203–1208. doi:10.1063/1.322842

Butzen PF (2012) Aging aware design techniques and CMOS gate degradation estimative. Ph.D. thesis (doctorate in microelectronics), Universidade Federal do Rio Grande do Sul (UFRGS), Porto Alegre, RS - Brazil

Cadence (2013) Cadence SOC encounter user guide. Available at: http://www.cadence.com/products/di/first_encounter/pages/default.aspx. Visited on: Jul 2013

Cadence (2015) Virtuoso layout suite for electrically aware design. Available at: http://www.cadence.com/products/cic/electrically_aware_design/pages/default.aspx. Visited on: Mar 2013

Cadence (2016) Cadence virtuoso liberate characterization solution. Available at: https://www.cadence.com/content/cadence-www/global/en_US/home/tools/customicanalogrfdesign/librarycharacterization/virtuosoliberatecharacterization.html

Chatterjee S, Fawaz M, Najm FN (2013) Redundancy-aware electromigration checking for mesh power grids. In: IEEE/ACM international conference on computer-aided design, ICCAD 2013. IEEE Press, Piscataway, NJ, pp 540–547

Chen W (1999) The VLSI handbook. Electrical engineering handbook. Taylor and Francis, New York. http://books.google.com.br/books?id=0r5LihlMogkC

Cheng YL, Lee SY, Chiu C, Wu K (2008) Back stress model on electromigration lifetime prediction in short length copper interconnects. In: IEEE international reliability physics symposium, IRPS 2008, pp 685–686. doi:10.1109/RELPHY.2008.4558988

Cheng Y, Todri-Sanial A, Bosio A, Dillio L, Girard P, Virazel A, Vevet P, Belleville M (2013) A novel method to mitigate tsv electromigration for 3d ics. In: IEEE computer society annual symposium on VLSI, ISVLSI 2013, pp 121–126. doi:10.1109/ISVLSI.2013.6654633

Choi ZS, Gan C, Wei F, Thompson CV, Lee J, Pey K, Choi W (2004) Fatal void size comparisons in via-below and via-above cu dual-damascene interconnects. In: MRS proceedings materials, technology and reliability of advanced interconnects, vol 812. Cambridge University Press, Cambridge, pp F7–6. doi:http://dx.doi.org/10.1557/PROC-812-F7.6

Domae S, Ueda T (2001) CMOS inverter and standard cell using the same. US Patent 6,252,427

Fawaz M, Chatterjee S, Najm FN (2013) A vectorless framework for power grid electromigration checking. In: International conference on computer-aided design, ICCAD 2013. IEEE Press, Piscataway, NJ, pp 553–560

Flach G, Reimann T, Posser G, Johann M, Reis R (2013) Simultaneous gate sizing and Vth assignment using lagrangian relaxation and delay sensitivities. In: IEEE computer society annual symposium on VLSI, ISVLSI 2013, IEEE, pp 84–89

Flach G, Reimann T, Posser G, Johann M, Reis R (2014) Effective method for simultaneous gate sizing and Vth assignment using lagrangian relaxation. IEEE Trans Comput Aided Des Integr Circuits Syst 33(4):546–557. doi:10.1109/TCAD.2014.2305847

FreePDK45 (2011) FreePDK45 process design kit. Available at: http://www.eda.ncsu.edu/wiki/FreePDK45:Contents. Visited on: Mar 2013

FreePDK15 (2014) FreePDK15 process design kit. Available at: http://www.eda.ncsu.edu/wiki/FreePDK15:Contents. Visited on: Nov 2014

Geden B (2011) Understand and avoid electromigration (em) & ir-drop in custom ip blocks. http://www.synopsys.com/Tools/Verification/CapsuleModule/CustomSim-RA-wp.pdf

Hu CK et al (2007) Impact of Cu microstructure on electromigration reliability. In: IEEE international interconnect technology conference, IITC 2007, pp 93–95

ITRS (2011) International technology roadmap for semiconductors. Available at: http://www.itrs.net/reports.html. Visited on: Mar 2014

Jain P, Jain A (2012) Accurate current estimation for interconnect reliability analysis. IEEE Trans Very Large Scale Integr (VLSI) Syst 20(9):1634–1644

Jain P, Cortadella J, Sapatnekar SS (2016) A fast and retargetable framework for logic-ip-internal electromigration assessment comprehending advanced waveform effects. IEEE Trans Very Large Scale Integr (VLSI) Syst 24(6):2345–2358. doi:10.1109/TVLSI.2015.2505504

Jerke G, Lienig J (2010) Early-stage determination of current-density criticality in interconnects. In: 11th international symposium on quality electronic design, ISQED 2010, pp 667–674. doi:10.1109/ISQED.2010.5450505

Jonggook K, Tyree V, Crowell C (1999) Temperature gradient effects in electromigration using an extended transition probability model and temperature gradient free tests. I. transition probability model. In: IEEE international integrated reliability workshop, IIRW 1999, pp 24–40. doi:10.1109/IRWS.1999.830555

Kahng A (2011) VLSI physical design: from graph partitioning to timing closure. Springer Science and Business Media, New York. http://books.google.com.br/books?id=DWUGHyFVpboC

Kahng A, Nath S, Rosing T (2013a) On potential design impacts of electromigration awareness. In: 18th Asia and South Pacific design automation conference, (ASP-DAC) 2013, pp 527–532. doi:10.1109/ASPDAC.2013.6509650

Kahng AB, Kang S, Lee H (2013b) Smart non-default routing for clock power reduction. In: 50th annual design automation conference, DAC 2013. ACM, New York, pp 91:1–91:7. doi:10.1145/2463209.2488846. http://doi.acm.org/10.1145/2463209.2488846

Kludt J, Weide-Zaage K, Ackermann M, Kovacs C, Hein V (2014) Reliability performance of different layouts of wide metal tracks. In: IEEE international reliability physics symposium, IRPS 2014, pp IT.4.1–IT.4.4. doi:10.1109/IRPS.2014.6861153

Lee JH (2012a) Implications of modern semiconductor technologies on gate sizing. PhD thesis, University of California, Los Angeles

Lee KD (2012b) Electromigration recovery and short lead effect under bipolar- and unipolar-pulse current. In: IEEE international reliability physics symposium, IRPS 2012, pp 6.B.3.1–6.B.3.4

Li B, Gill J, Christiansen C, Sullivan T, McLaughlin PS (2005) Impact of via-line contact on Cu interconnect Electromigration performance. In: IEEE international reliability physics symposium, IRPS 2005, pp 24–30. doi:10.1109/RELPHY.2005.1493056

Li B, Christiansen C, Badami D, Yang CC (2014) Electromigration challenges for advanced on-chip cu interconnects. Microelectron Reliab 54(4):712–724. doi:http://dx.doi.org/10.1016/j.microrel.2014.01.005. http://www.sciencedirect.com/science/article/pii/S0026271414000092

Li Da, Marek-Sadowska M, Nassif S (2015a) Layout aware electromigration analysis of power/ground networks. In: Reis R, Cao Y, Wirth G (eds) Circuit design for reliability. Springer, New York, pp 145–173. doi:10.1007/978-1-4614-4078-9_8. http://dx.doi.org/10.1007/978-1-4614-4078-9_8

Li DA, Marek-Sadowska M, Nassif SR (2015b) T-vema: A temperature- and variation-aware electromigration power grid analysis tool. IEEE Trans Very Large Scale Integr (VLSI) Syst 23(10):2327–2331. doi:10.1109/TVLSI.2014.2358678

Lienig J (2006) Introduction to electromigration-aware physical design. In: International symposium on physical design, ISPD 2006. ACM, New York, pp 39–46. doi:10.1145/1123008.1123017. http://doi.acm.org/10.1145/1123008.1123017

Lienig J (2013) Electromigration and its impact on physical design in future technologies. In: ACM international symposium on physical design, ISPD 2013, pp 33–40

Liu Y, Li M, Kim DW, Gu S, Tu KN (2015) Synergistic effect of electromigration and joule heating on system level weak-link failure in 2.5D integrated circuits. J Appl Phys 118(13):135304. doi:http://dx.doi.org/10.1063/1.4932598. http://scitation.aip.org/content/aip/journal/jap/118/13/10.1063/1.4932598

Maricau E, Gielen G (2013) CMOS reliability overview. In: Analog IC reliability in nanometer CMOS, analog circuits and signal processing. Springer, New York, pp 15–35. doi:10.1007/978-1-4614-6163-0_2. http://dx.doi.org/10.1007/978-1-4614-6163-0_2

Mentor (2013) Calibre xRC. Available at: http://www.mentor.com. Visited on: Mar 2013

Mishra V, Sapatnekar S (2013) The impact of electromigration in copper interconnects on power grid integrity. In: 50th ACM/IEEE design automation conference, DAC 2013, pp 1–6

Mishra V, Sapatnekar SS (2015) Circuit delay variability due to wire resistance evolution under ac electromigration. In: 2015 IEEE international reliability physics symposium, pp 3D.3.1–3D.3.7. doi:10.1109/IRPS.2015.7112713

Mishra V, Sapatnekar SS (2016) Predicting electromigration mortality under temperature and product lifetime specifications. In: Proceedings of the 53rd annual design automation conference, DAC '16. ACM, New York, pp 43:1–43:6. doi:10.1145/2897937.2898070. http://doi.acm.org/10.1145/2897937.2898070

Nangate (2011) Nangate open cell library v1.0, FreePDK v1.3 package. Available at: http://www.nangate.com. Visited on: Sept 2013

Nangate (2014) Nangate FreePDK15 open cell library v0.1. Available at: http://www.nangate.com. Visited on: Jun 2014

Nastase AS (2013) How to derive the RMS value of a triangle waveform. Available at: http://masteringelectronicsdesign.com/how-to-derive-the-rms-value-of-a-triangle-waveform. Visited on: Apr 2013

Nguyen H, Salm C, Wenzel R, Mouthaan A, Kuper F (2002) Simulation and experimental characterization of reservoir and via layout effects on electromigration lifetime. Microelectron Reliab 42(9–11):1421–1425. http://doc.utwente.nl/67753/

Pak J, Lim SK, Pan DZ (2013) Electromigration study for multi-scale power/ground vias in tsv-based 3d ics. In: International conference on computer-aided design, ICCAD 2013. IEEE Press, Piscataway, NJ, pp 379–386

Park YJ, Jain P, Krishnan S (2010) New electromigration validation: via node vector method. In: International reliability physics symposium, IRPS 2010, pp 698–704

Patel SJ (2014) Problems identification and proposed solutions in asic physical designing. Program Device Circuits Syst 6(3):89–91

Pelloie JL (2013) Method of adapting a layout of a standard cell of an integrated circuit. US Patent 8,381,162

Posser G, Flach G, Wilke G, Reis R (2012) Gate sizing using geometric programming. Analog Integr Circuits Sig Process 73(3):831–840

Posser G, Belomo J, Meinhardt C, Reis R (2014a) Perfomance improvement with dedicated transistor sizing for mosfet and finfet devices. In: IEEE computer society annual symposium on VLSI, ISVLSI 2014, IEEE, pp 418–423

Posser G, Mishra V, Jain P, Reis R, Sapatnekar SS (2014b) A systematic approach for analyzing and optimizing cell-internal signal electromigration. In: IEEE/ACM international conference on computer-aided design, ICCAD 2014. IEEE Press, Piscataway, NJ, pp 486–491. http://dl.acm.org/citation.cfm?id=2691365.2691463

Posser G, Mishra V, Reis R, Sapatnekar SS (2014c) Analyzing the electromigration effects on different metal layers and different wire lengths. In: 21st IEEE international conference on electronics, circuits and systems, ICECS 2014, Marseille, pp 682–685

Pullela S, Menezes N, Pillage L (1995) Low power ic clock tree design. In: IEEE custom integrated circuits conference, CICC 1995, pp 263–266. doi:10.1109/CICC.1995.518182

Rabaey JM, Chandrakasan AP, Nikolic B (2002) Digital integrated circuits. Prentice Hall, Englewood Cliffs

Reimann T, Posser G, Flach G, Johann M, Reis R (2013) Simultaneous gate sizing and Vt assignment using fanin/fanout ratio and Simulated Annealing. In: IEEE international symposium on circuits and systems, ISCAS 2013, IEEE, pp 2549–2552

Reis R, Cao Y, Wirth G (2015) Circuit design for reliability. Springer, New York

Sapatnekar S (2013) What happens when circuits grow old: aging issues in CMOS design. In: International symposium on VLSI Technology, systems, and applications, VLSI-TSA 2013, pp 1–2. doi:10.1109/VLSI-TSA.2013.6545621

Sengupta D, Sapatnekar SS (2014) Rescale: recalibrating sensor circuits for aging and lifetime estimation under bti. In: IEEE/ACM international conference on computer-aided design, ICCAD 2014. IEEE Press, Piscataway, pp 492–497

SI2 (2009) LEF DEF guide. Available at: http://www.si2.org/openeda.si2.org/projects/lefdef. Visited on: Apr 2013

Skadron K, Stan M, Huang W, Velusamy S, Sankaranarayanan K, Tarjan D (2003) Temperature-aware microarchitecture. In: 30th annual international symposium on computer architecture, ISCA 2003, pp 2–13. doi:10.1109/ISCA.2003.1206984

Srinivasan J, Adve S, Bose P, Rivers J (2004) The impact of technology scaling on lifetime reliability. In: International conference on dependable systems and networks, DSN 2004, pp 177–186

Summers K (2013) Five-minute tutorial: creating an em model file. Available at: http://www.cadence.com/Community/blogs/di/archive/2013/01/14/five-minute-tutorial-creating-an-em-model-file.aspx. Visited on: Jan 2014

Synopsys (2013a) Cosmosscope: premier graphical waveform analyzer. Available at: http://www.synopsys.com/Prototyping/Saber/Pages/cosmos_scope_ds.aspx. Visited on: Mar 2013

Synopsys (2013b) Synopsys design compiler user guide. Available at: http://www.synopsys.com/Tools/Implementation/RTLSynthesis/DCUltra/pages/default.aspx. Visited on: Jun 2013

Synopsys (2014a) Ic compiler: comprehensive place and route system. Http://www.synopsys.com/Tools/Implementation/PhysicalImplementation/Documents/iccompiler_ds.pdf.   Visited on: May 2013

Synopsys (2014b) Synopsys 32/28nm open pdk. Available at: http://www.synopsys.com. Visited on: Mar 2014

Synopsys (2016) Synopsys siliconsmart. Available at: http://www.synopsys.com/Tools/Implementation/SignOff/Pages/siliconsmartds.aspx.

Thompson C (2008) Using line-length effects to optimize circuit-level reliability. In: 15th international symposium on the physical and failure analysis of integrated circuits, 2008. IPFA 2008, IEEE, pp 1–4

Ting L, May J, Hunter W, McPherson J (1993) AC electromigration characterization and modeling of multilayered interconnects. In: International reliability physics symposium, IRPS 1993, IEEE, pp 311–316

Tu KN (2003) Recent advances on electromigration in very-large-scale-integration of interconnects. J Appl Phys 94(9):5451–5473. doi:http://dx.doi.org/10.1063/1.1611263. http://scitation.aip.org/content/aip/journal/jap/94/9/10.1063/1.1611263

Tu K (2011) Reliability challenges in 3D IC packaging technology. Microelectron Reliab 51(3):517–523. doi:http://dx.doi.org/10.1016/j.microrel.2010.09.031. http://www.sciencedirect.com/science/article/pii/S0026271410005214

Vaidyanathan K, Liebmann L, Strojwas A, Pileggi L (2014a) Sub-20nm design technology co-optimization for standard cell logic. In: IEEE/ACM international conference on computer-aided design, ICCAD 2014. IEEE Press, Piscataway, NJ, pp 124–131. http://dl.acm.org/citation.cfm?id=2691365.2691393

Vaidyanathan K, Liu R, Liebmann L, Lai K, Strojwas AJ, Pileggi L (2014b) Design implications of extremely restricted patterning. J Micro/Nanolithogr MEMS MOEMS 13(3):031309–031309

Wang CH, Tam KH, Chen HY (2014) Automatic place and route method for electromigration tolerant power distribution. US Patent 8,694,945

Weste NHE, Harris D (2005) CMOS VLSI design: a circuits and systems perspective. Addison-Wesley Publishing Company, Boston

White D, McSherry M, Fischer E, Yanagida B, Gopalakrishnan P (2016) Methods, systems, and articles of manufacture for implementing electronic circuit designs with electro-migration awareness. US Patent 9,330,222. https://www.google.com/patents/US9330222

Wirth G, da Silva R (2010) Noise and aging effects. Lecture at first IEEE cass summer school. http://inf.ufrgs.br/cass-school/

Wu KC, Lee MC, Marculescu D, Chang SC (2012) Mitigating lifetime underestimation: a system-level approach considering temperature variations and correlations between failure mechanisms. In: Design, automation test in Europe conference exhibition, DATE 2012, pp 1269–1274. doi:10.1109/DATE.2012.6176687

Xie J, Narayanan V, Xie Y (2012) Mitigating electromigration of power supply networks using bidirectional current stress. In: Great lakes symposium on VLSI, GLSVLSI 2012, pp 299–302

Yeric G, Cline B, Sinha S, Pietromonaco D, Chandra V, Aitken R (2013) The past present and future of design-technology co-optimization. In: IEEE custom integrated circuits conference, CICC 2013, IEEE, pp 1–8

Zhao W, Cao Y (2007) Predictive technology model for nano-cmos design exploration. J Emerg Technol Comput Syst 3(1):1. doi:http://doi.acm.org/10.1145/1229175.1229176

Zhao X, Wan Y, Scheuermann M, Lim SK (2013) Transient modeling of tsv-wire electromigration and lifetime analysis of power distribution network for 3d ICS. In: International conference on computer-aided design, ICCAD 2013. IEEE Press, Piscataway, NJ, pp 363–370

Zhou C, Wang X, Fung R, Wen SJ, Wong R, Kim CH (2015) High frequency AC electromigration lifetime measurements from a 32nm test chip. In: 2015 symposium on VLSI technology (VLSI technology), pp T42–T43. doi:10.1109/VLSIT.2015.7223696